T0235747

Lecture Notes in Computer Science   12473

More information about this series at http://www.springer.com/series/7409

Sergio Di Martino · Zhixiang Fang ·
Ki-Joune Li (Eds.)

Web and Wireless Geographical Information Systems

18th International Symposium, W2GIS 2020
Wuhan, China, November 13–14, 2020
Proceedings

 Springer

Editors
Sergio Di Martino ⓘ
University of Naples Federico II
Naples, Italy

Ki-Joune Li ⓘ
Department of Computer Science
and Engineering
Pusan National University
Pusan, Korea (Republic of)

Zhixiang Fang ⓘ
State Key Laboratory of Information
Engineering in Surveying, Mapping
and Remote Sensing
Wuhan University
Wuhan, China

ISSN 0302-9743 ISSN 1611-3349 (electronic)
Lecture Notes in Computer Science
ISBN 978-3-030-60951-1 ISBN 978-3-030-60952-8 (eBook)
https://doi.org/10.1007/978-3-030-60952-8

LNCS Sublibrary: SL3 – Information Systems and Applications, incl. Internet/Web, and HCI

This Springer imprint is published by the registered company Springer Nature Switzerland AG
The registered company address is: Gewerbestrasse 11, 6330 Cham, Switzerland

Preface

This volume contains the papers selected for presentation at the 18th edition of the International Symposium on Web and Wireless Geographical Information Systems (W2GIS 2020), hosted by Wuhan University, China, during November 13–14, 2020.

W2GIS is a series of events alternating between Europe and East Asia. It aims at providing a forum for discussing advances in theoretical, technical, and practical issues in the field of wireless and Internet technologies, suited for the spreading, usage, and processing of geo-referenced data. W2GIS now represents a prestigious event within the research community that continues to develop and expand. For the 2020 edition, we received 40 submissions from 12 countries on 4 continents. Each paper received three reviews and based on these reviews, 8 full papers and 15 progress papers or short papers were selected for presentation at the symposium and inclusion in this Springer LNCS volume (12473). The accepted papers are all of excellent quality and cover topics that range from mobile GIS and Location-Based Services to Spatial Information Retrieval and Wireless Sensor Networks.

We had keynote speeches by Dr. Christophe Claramunt from the Naval Academy Research Institute (France), Dr. Shih-Lung Shaw from The University of Tennessee, Knoxville (USA), and Dr. Yu Liu from Peking University (China).

We wish to thank all authors that contributed to this symposium for the high quality of their papers and presentations. Our sincere thanks go to Springer's LNCS team. We would also like to acknowledge and thank the Program Committee members for the quality and timeliness of their reviews. Finally, many thanks to Christophe Claramunt and the entire Steering Committee for providing continuous support and advice.

August 2020
Sergio Di Martino
Zhixiang Fang
Ki-Joune Li

Organization

Program Committee

Masatoshi Arikawa	The University of Tokyo, Japan
Andrea Ballatore	Birkbeck, University of London, UK
Michela Bertolotto	University College Dublin, Ireland
Alain Bouju	University of La Rochelle, France
Christophe Claramunt	Naval Academy Research Institute, France
Maria Luisa Damiani	University of Milan, Italy
Sergio Di Martino	University of Naples Federico II, Italy
Matt Duckham	RMIT University, Australia
Zhixiang Fang	Wuhan University, China
Filomena Ferrucci	Università di Salerno, Italy
Stefan Funke	University of Stuttgart, Germany
Jerome Gensel	Laboratoire d'Informatique de Grenoble, France
Haosheng Huang	Ghent University, Belgium
Kyoung-Sook Kim	National Institute of Advanced Industrial Science and Technology (AIST), Japan
Daisuke Kitayama	Kogakuin University, Japan
Songnian Li	Ryerson University, Canada
Xiang Li	East China Normal University, China
Feng Lu	The University of Tokyo, Japan
Miguel R. Luaces	Universidade da Coruña, Spain
Miguel Mata	UPIITA-IPN, Mexico
Gavin McArdle	Maynooth University, Ireland
Kostas Patroumpas	Information Management Systems Institute, Athena Research Center, Greece
Shiori Sasaki	Keio University, Japan
Sabine Storandt	University of Konstanz, Germany
Taro Tezuka	University of Tsukuba, Japan
Yuanyuan Wang	Yamaguchi University, Japan
Yousuke Watanabe	Nagoya University, Japan
Robert Weibel	University of Zurich, Switzerland
Stephan Winter	The University of Melbourne, Australia
Junjun Yin	Penn State University, USA
F. Javier Zarazaga-Soria	University of Zaragoza, Spain
Zhuqing Zhu	Wuhan University, China

Contents

Crowdsourcing, Volunteered Geographic Information and Social Networks

Spatial Algorithms

Indoor/Outdoor Localization and Navigation

How to Extend IndoorGML for Points of Interest
[Work-in-Progress Paper]

Taehoon Kim[1,3], Kyoung-Sook Kim[1(✉)], Jiyeong Lee[2], and Ki-Joune Li[3]

[1] National Institute of Advanced Industrial Science and Technology (AIST),
Tokyo, Japan
{kim.taehoon,ks.kim}@aist.go.jp
[2] The University of Seoul, Seoul, Republic of Korea
jlee@uos.ac.kr
[3] Pusan National University, Pusan, Republic of Korea
{taehoon.kim,lik}@pusan.ac.kr

Abstract. Nowadays, the interest in spatial information services is growing and moving towards indoor spaces. Indoor spatial data is fundamental as the demand for expressing complex urban environments, in the context of providing location-based services. OGC IndoorGML was established as a geospatial data standard, especially for indoor navigation. OGC IndoorGML provides a broad definition of the expression and structure of indoor spaces, but point of interest (POI) is excluded from their scope. However, POIs are useful for indoor navigation not only outdoor navigation. In this paper, we propose a data model that describes POIs in an indoor environment considering various spatial and temporal aspects, as an extension model of the OGC IndoorGML, called indoor POI (InPOI) model. The InPOI model defines new entities to handle Spatio-temporal POI information in indoor space by specifying the OGC IndoorGML core module. Finally, we provide an XML schema of the InPOI model.

Keywords: OGC IndoorGML · Indoor POI · XML Schema

1 Introduction

Nowadays, day-to-day human activities have been closely tied with mobile devices and gadgets that continuously improving in terms of features and speeds while decreasing in size, mostly equipped with GPS receivers and cameras [9,10]. With this, Location-Based Services (LBS) deliver relevant and timely information to mobile users based on their positions [3,9]. These services form a part of the core infrastructures of smart cities, as localities around the world aim to establish seamless integration of technology to the daily life of its citizens.

Points of Interest (POI) data, also sometimes cited as "Places of Interest [1]", "Objects of Interest [2]", or "Landmarks [12]" in some literature, are one of

© Springer Nature Switzerland AG 2020
S. Di Martino et al. (Eds.): W2GIS 2020, LNCS 12473, pp. 3–13, 2020.
https://doi.org/10.1007/978-3-030-60952-8_1

the fundamental requirement of any geospatial data infrastructure. They usually point out labeling features in maps instead of geographic coordinates of objects [10]. They are used in a wide range of geospatial applications, such as tourist guidance, indoor mapping [11], 3D visualization [13], and so on. Also, we can define new POI depending on the requirements of the application. In recent, indoor space has become a fundamental domain of geospatial applications and services. Like restaurants in outdoor, POI in indoor, such as toilets and vending machines, can help mobile users to improve their geospatial cognition. However, most indoor LBS are still providing viewer-level service [6]. To support various indoor applications, including navigation, simulation, monitoring, and even facility management, POI is required to be defined as indoor geospatial information.

Indoor Geographic Markup Language (IndoorGML [8]) is the Open Geospatial Consortium (OGC) standard for indoor geospatial information, particularly for navigation systems. OGC IndoorGML focuses on how topological relationships would be represented in indoor space. The OGC IndoorGML provides a broad definition of the expression and structure of indoor spaces, but POIs (such as event, facility, furniture, and installation) are out of their scope. However, the absence of POI information in indoor space brings the limitation of standard implementation and usability.

In this paper, we propose a data model that describes POI in the indoor environment, considering its various spatial and temporal aspects. This paper consists of three parts as follows: In Sect. 2, we briefly introduce previous POI standardization studies and OGC IndoorGML. In the next section, we present our proposed indoor POI model based on OGC IndoorGML, concludes with plans for further developments.

2   Related Studies

2.1   Point of Interest (POI)

POI data is one of the most fundamental requirements in the core infrastructure of smart cities. The definition of POI which provided by the OGC is "A location (with a fixed position) where one can find a place, product or service, typically identified by name rather than by address and characterized by type, which may be used as a reference point or a target in a location based service request." [7] Internationally, there have been activities relating to POIs:

- ISO 19112 [4] connects location references to coordinates. It defined the components of a spatial reference system and the essential components of a gazetteer.
- ISO 19155 [5] defined mechanisms to match multiple Place Identifiers (PI) to the same place, in which an identifier of a place is referred to as a PI.
- The World Wide Web Consortium (W3C) POI Special Working Group (SWG) has become the first step towards the standardization for the definition and exchange of POI data, focusing on web architecture and in AR

applications. In parallel, industry stakeholders, including car industry special-
ists and experts in mobile technology, navigation systems, and digital maps
formulated a general-purpose specification called Point of Interest Exchange
Language (POIX) and submitted this as a preliminary proposal to W3C.
However, this lacked in certain, descriptive and temporal aspects, and seemed
to be geared towards car navigation. The W3C POI SWG has published the
Point of Interest Core (POI Core), describing eight categories to describe POI
with attributes and various location types. However, it has been described as
"quite complex and requires substantial effort to apply."
– Since 2015, the OGC officially started activities for this purpose in its POI
SWG. It has since issued OpenPOIs, a public database of POI data based on
the earlier W3C model. This, however, contains a large amount of metadata
that significantly increases data volume. As of writing, OGC is still undergo-
ing the standardization process.

One major problem that having a POI data standard would resolve is having separate sets of POIs for every application, or infrastructure. For numerous
sources of POI data that are available, the W3C workshops raised the issue of
connecting POI data from different sources, avoiding information overload, and
identifying the veracity of information. For example, in Korea, of the 3 Million
POIs being maintained by 12 private companies, 70% have overlapped. For POI
data to be comfortable to distribute and lightweight, it must be general-purpose
and is dynamically extensible.

2.2   OGC IndoorGML

OGC IndoorGML [8] was established as the OGC international standard for
indoor navigation applications and XML-based formats to represent indoor spatial information.

OGC IndoorGML provides a standard data model for indoor space with
two spatial models, as shown in Fig. 1: Euclidean Space represents the shape
of a three-dimensional (3D) cell space; Topology Space represents connectivity
between cell spaces. Topology represents a duality transformation of the 3D cell
space and is an essential component for indoor navigation and routing system.
By applying a duality transformation, the 3D cells in primal space are mapped
to nodes (0D) in dual space. The topological adjacency relationships between 3D
cells are transformed to edges (1D) linking pairs of nodes in dual space. Therefore, OGC IndoorGML utilizes a network model for navigation and expresses
the connectivity relationships among cell spaces. The nodes of the indoor network represent rooms, corridors, doors, elevators, and staircases. The edges of
the indoor network represent the topological relationships among indoor spatial entities and can indicate the paths of pedestrian movement between nodes
within a building. Therefore, one edge should be represented by two nodes. The
network model in OGC IndoorGML is represented by nodes (as called **State**)
and edges (as called **Transition**) feature, as shown in Fig. 2.

However, in certain applications, OGC IndoorGML would necessitate information from other datasets to provide comprehensive information especially in LBS. Indoor POIs provide a useful location on objects, events, or facilities in a specific location in the indoor environment, and may be utilized together with OGC IndoorGML for a more complete depiction of indoor space.

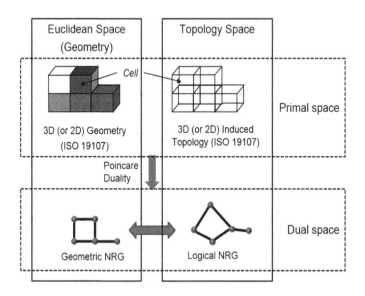

Fig. 1. Structure space model (OGC, IndoorGML [8])

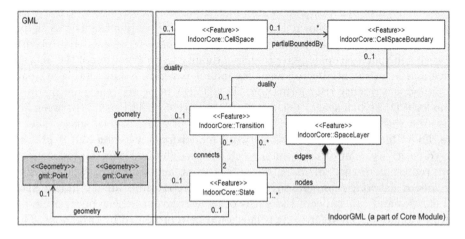

Fig. 2. Part of OGC IndoorGML Core module UML diagram (OGC, IndoorGML [8])

3   Indoor POI Data Model

The indoor POI data model aims to supplement existing descriptions of POI for a more suitable standard in creating such features in the indoor space, to increase utilization and development of indoor location-based services. The data model expresses objects and their respective logical relationships through the Unified Modeling Language (UML).

3.1   Indoor POI Classification

To encourage utilization, increase efficiency and avoid duplication in datasets, a classification scheme was devised to categorize Indoor POI. Created based on the ESRI POI classification scheme, each feature may be classified in three levels of increasing specificity, and each category would correspond to a 6-character category code as an attribute for the indoor POI. Figure 3 shows an example of the classification scheme for indoor POIs.

Indoor POI (reclassified)								
Level 1	Level 2	Level 3	Level 1	Level 2	Level 3	Level 1	Level 2	Level 3
Place	Pedestrian	stairs	Retail + Services	Services	nursery	Safety + Security	Fire-fighting supplies	fire protection appliance
		slope way			drug store			fire extinguishing system
		lobby			vending machine			fire extinguisher
	Private	common room			ticket machine			fire alarm
		relaxation room			information		Emergency relief supplies	automated external defibrillator
	Relaxation	smoking area			lounge			life-saving trolley
		men's toilet			guest room			shelter
		woman's toilet			covered car park			emergency call center
		disabled toilet			bank		Evacuation facility	exit
		bench			cash machine or ATM			emergency escape device
		rubbish bin			post office			relief goods
Things	Access facility	door			billboard	Event	Temporal	time
		ticket gate		Retail	restaurant		Type	fire
		access control units			coffee shop			earthquake
	Conveyor transport	up escalator			clothing store			power outage
		down escalator			hair shop			safety accident
		up moving sidewalk			ticket office			
		down moving sidewalk			aquarium			
		Horizontal moving sidewalk			bowling alley			
		Elevator			swimming-pool			
		Wheelchair lift			store			

Fig. 3. 3-level indoor POI classification schema

3.2   Indoor POI Model

The proposed indoor POI (InPOI) Model is shown in Fig. 4. The InPOI model consists of five elements: **IndoorPOISpace, IndoorPOILocation, Indoor-POIAttribute, IndoorPOI**, and **IndoorPOIs**. The UML diagram depicted in Fig. 4 shows the InPOI data model based on the OGC IndoorGML core module. The main class of the InPOI module is the **IndoorPOI** class. The **IndoorPOI** class can have one or more **IndoorPOILocation** class instances, at least. POI location information is mandatory for managing POI history. Optionally, the

IndoorPOI class can have multiple **IndoorPOISpace** and **IndoorPOIAttribute** instance. **IndoorPOISpace** instance represents the occupied space of POI. It can have a duality attribute that inherits from the **CellSpace** of OGC IndoorGML core module. The duality attribute of **IndoorPOISpace** class represents an association with **IndoorPOILocation**. The **IndoorPOILocation** also can have a duality attribute for representing an association with **IndoorPOISpace**. The detail description of each class is as below:

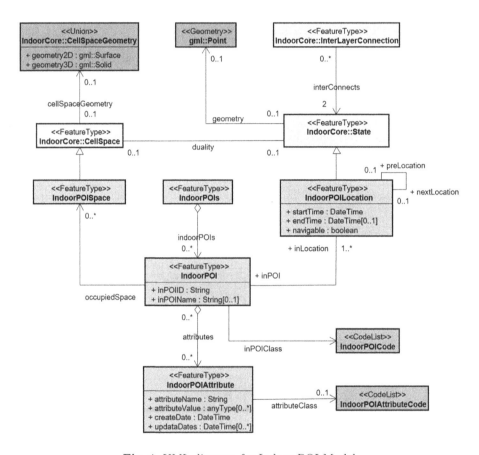

Fig. 4. UML diagram for Indoor POI Model

⟨IndoorPOISpace⟩
IndoorPOISpace represents an occupied space of POI. The XML schema for **IndoorPOISpace** as shown in Fig. 5. **IndoorPOISpace** class inherits **CellSpace** class, therefore, **IndoorPOISpace** has a geometry that derived from **CellSpace** class. This geometry information can be useful for creating dynamic map information. For example, to create a map for an autonomous robot, it is important whether it is enough space for the robot to pass through. At this time, this geometry information can be utilized.

```
<xs:element name="IndoorPOISpace" type="IndoorPOISpaceType" substitutionGroup="IndoorCore:CellSpace"/>
<!-- =================================================================== -->
<xs:complexType name="IndoorPOISpacePropertyType">
  <xs:sequence minOccurs="0">
    <xs:element ref="IndoorPOISpace"/>
  </xs:sequence>
  <xs:attributeGroup ref="gml:AssociationAttributeGroup"/>
</xs:complexType>
<!-- =================================================================== -->
<xs:complexType name="IndoorPOISpaceType">
  <xs:complexContent>
    <xs:extension base="IndoorCore:CellSpaceType">
      <xs:attributeGroup ref="gml:AssociationAttributeGroup"/>
    </xs:extension>
  </xs:complexContent>
</xs:complexType>
```

Fig. 5. XML Schema for **IndoorPOISpace**

⟨**IndoorPOILocation**⟩

IndoorPOILocation represents a specific point location of POI. The XML schema for **IndoorPOILocation** is shown in Fig. 6. **IndoorPOILocation** has a geometry that derived from **State** class, and it can express *WITHIN* topological relation with **CellSpace** using **InterLayerConnection**; i.e., All POIs within a room have *interConnects* with same **State** of a room. The *startTime* and *endTime* elements are installed and uninstalled time of POI respectively. The *navigable* element indicates which people can movable through this POI or not; i.e., usually, the vending machine is not navigable, but the ticket gate is navigable. This element can be used for making an indoor route. **IndoorPOILocation** contains the *preLocation* and *nextLocation* for tracking the location history

```
<xs:element name="IndoorPOILocation" type="IndoorPOILocationType" substitutionGroup="IndoorCore:State"/>
<!-- =================================================================== -->
<xs:complexType name="IndoorPOILocationPropertyType">
  <xs:sequence minOccurs="0">
    <xs:element ref="IndoorPOILocation"/>
  </xs:sequence>
  <xs:attributeGroup ref="gml:AssociationAttributeGroup"/>
</xs:complexType>
<!-- =================================================================== -->
<xs:complexType name="IndoorPOILocationType">
  <xs:complexContent>
    <xs:extension base="IndoorCore:StateType">
      <xs:sequence>
        <xs:element name="startTime" type="xs:dateTime"/>
        <xs:element name="endTime" type="xs:dateTime" minOccurs="0"/>
        <xs:element name="navigable" type="xs:boolean"/>
        <xs:element name="preLocation" type="IndoorPOILocationPropertyType" minOccurs="0"/>
        <xs:element name="nextLocation" type="IndoorPOILocationPropertyType" minOccurs="0"/>
        <xs:element name="inPOI" type="IndoorPOIPropertyType"/>
      </xs:sequence>
      <xs:attributeGroup ref="gml:AssociationAttributeGroup"/>
    </xs:extension>
  </xs:complexContent>
</xs:complexType>
```

Fig. 6. XML Schema for **IndoorPOILocation**

of POI, when a POI move to another location. Lastly, the *inPOI* element represents an association with the corresponding **IndoorPOI** class which represents a POI information.

⟨**IndoorPOIAttribute**⟩
IndoorPOIAttribute is used to describe the indoor POI property. The XML schema for **IndoorPOIAttribute** as shown in Fig. 7. The *attributeName* and *attributeValue* elements are name and value of user-defined attribute, respectively. The *createDate* and *updateDate* elements are created and updated date of attribute, respectively. The *attributeClass* element is a flexible enumeration that uses string values for expressing a list of attribute class code.

```
<xs:element name="IndoorPOIAttribute" type="IndoorPOIAttributeType" substitutionGroup="gml:AbstractFeature"/>
<!-- ================================================================================ -->
<xs:complexType name="IndoorPOIAttributePropertyType">
  <xs:sequence minOccurs="0">
    <xs:element ref="IndoorPOIAttribute"/>
  </xs:sequence>
  <xs:attributeGroup ref="gml:AssociationAttributeGroup"/>
</xs:complexType>
<!-- ================================================================================ -->
<xs:complexType name="IndoorPOIAttributeType">
  <xs:complexContent>
    <xs:extension base="gml:AbstractFeatureType">
      <xs:sequence>
        <xs:element name="attributeName" type="xs:string"/>
        <xs:element name="attributeValue" type="xs:anyType" minOccurs="0" maxOccurs="unbounded"/>
        <xs:element name="createDate" type="xs:dateTime"/>
        <xs:element name="updateDates" type="xs:dateTime" minOccurs="0" maxOccurs="unbounded"/>
        <xs:element name="attributeClass" type="IndoorPOIAttributeCodeType" minOccurs="0" />
      </xs:sequence>
      <xs:attributeGroup ref="gml:AggregationAttributeGroup"/>
    </xs:extension>
  </xs:complexContent>
</xs:complexType>
<!-- ================================================================================ -->
<xs:simpleType name="IndoorPOIAttributeCodeType">
  <xs:restriction base="xs:string">
    <xs:enumeration value="..."/>
      ...
  </xs:restriction>
</xs:simpleType>
```

Fig. 7. XML Schema for **IndoorPOIAttribute**

⟨**IndoorPOI**⟩
IndoorPOI class is a feature class for representing indoor POI. The XML schema for **IndoorPOI** as shown in Fig. 8. To represent spatial objects of indoor POI in indoor space, an **IndoorPOI** has *occupiedSpace* element. Also, to represent location of indoor POI in indoor space, an **IndooPOI** has *inLocation* element. The **IndoorPOI** class has attributes which are *inPOIID*, *inPOIName* and *inPOIClass*. The *inPOIID* and *inPOIName* are the identifier and name of indoor POI, respectively. The attribute *inPOIClass* represents a general classification code of indoor POI, e.g., the classification scheme described at Sect. 3.1. An **IndoorPOI** aggregates **IndoorPOIAttribute** which is user-defined attribute.

```
<xs:element name="IndoorPOI" type="IndoorPOIType" substitutionGroup="gml:AbstractFeature"/>
<!-- ================================================================= -->
<xs:complexType name="IndoorPOIPropertyType">
  <xs:sequence minOccurs="0">
    <xs:element ref="IndoorPOI"/>
  </xs:sequence>
  <xs:attributeGroup ref="gml:AssociationAttributeGroup"/>
</xs:complexType>
<!-- ================================================================= -->
<xs:complexType name="IndoorPOIType">
  <xs:complexContent>
    <xs:extension base="gml:AbstractFeatureType">
      <xs:sequence>
        <xs:element name="inPOIID" type="xs:string"/>
        <xs:element name="inPOIName" type="xs:string" minOccurs="0"/>
        <xs:element name="inLocation" type="IndoorPOILocationPropertyType" maxOccurs="unbounded"/>
        <xs:element name="occupiedSpace" type="IndoorPOISpacePropertyType" minOccurs="0" maxOccurs="unbounded"/>
        <xs:element name="attributes" type="IndoorPOIAttributePropertyType" minOccurs="0" maxOccurs="unbounded"/>
        <xs:element name="inPOIClass" type="IndoorPOICodeType"/>
      </xs:sequence>
      <xs:attributeGroup ref="gml:AggregationAttributeGroup"/>
    </xs:extension>
  </xs:complexContent>
</xs:complexType>
<!-- ================================================================= -->
<xs:simpleType name="IndoorPOICodeType">
  <xs:restriction base="xs:string">
    <xs:enumeration value="..."/>
      ...
  </xs:restriction>
</xs:simpleType>
```

Fig. 8. XML Schema for **IndoorPOI**

⟨**IndoorPOIs**⟩
IndoorPOIs is a root element of to represent the indoor POI. It is an aggregated element with **IndoorPOI**. The **IndoorPOIs** contains a set of **IndoorPOI** as *indoorPOIs*. The XML schema for **IndoorPOI** as shown in Fig. 9.

```
<xs:element name="IndoorPOIs" type="IndoorPOIsType" substitutionGroup="gml:AbstractFeature"/>
<!-- ================================================================= -->
<xs:complexType name="IndoorPOIsType">
  <xs:complexContent>
    <xs:extension base="gml:AbstractFeatureType">
      <xs:sequence>
        <xs:element name="indoorPOIs" type="IndoorPOIPropertyType" minOccurs="0" maxOccurs="unbounded"/>
      </xs:sequence>
      <xs:attributeGroup ref="gml:AssociationAttributeGroup"/>
    </xs:extension>
  </xs:complexContent>
</xs:complexType>
```

Fig. 9. XML Schema for **IndoorPOIs**

4   Conclusion

OGC IndoorGML was an established OGC standard for indoor spatial information with a focus on navigation. Still, for specific applications, information from other datasets is needed to provide comprehensive information, especially in LBS. Indoor POIs provide a useful location for objects, events, or facilities at specific locations in an indoor environment, and can be used with OGC IndoorGML to more fully depict indoor spaces. This study presents a data model that describes POIs in indoor environments, taking into account various spatial and temporal aspects based on OGC IndoorGML.

However, further researches are required to verify the purposed extension model. Further research will include various use-case, for example, facility management in the hospital. And then, finally, we plan to submit the InPOI model to OGC as a discussion paper with use-cases.

Acknowledgment. This research was partially supported by the New Energy and Industrial Technology Development Organization (NEDO) and a grant(19NSIP-B135746-03) from National Spatial Information Research Program (NSIP) funded by Ministry of Land, Infrastructure and Transport of Korean government.

References

1. Baltrunas, Linas, Ludwig, Bernd, Peer, Stefan, Ricci, Francesco: Context-aware places of interest recommendations for mobile users. In: Marcus, Aaron (ed.) DUXU 2011. LNCS, vol. 6769, pp. 531–540. Springer, Heidelberg (2011). https://doi.org/10.1007/978-3-642-21675-6_61

2. De Carolis, B., Novielli, N., Plantamura, V.L., Gentile, E.: Generating comparative descriptions of places of interest in the tourism domain. In: Proceedings of the Third ACM Conference on Recommender Systems, pp. 277–280. ACM (2009)

3. Heikkinen, A., Okkonen, A., Karhu, A., Koskela, T.: A distributed POI data model based on the entity-component approach. In: 2014 IEEE Symposium on Computers and Communications (ISCC), pp. 1–6. IEEE (2014)

4. International Organization for Standardization: Geographic Information: Spatial Referencing by Geographic Identifiers. ISO 19112:2003 (2003)

5. International Organization for Standardization: Geographic Information: Geographic Information - Place Identifier (PI) Architecture. ISO 19155:2017 (2018)

6. Jung, H.J., Lee, J.: Development of an omnidirectional-image-based data model through extending the IndoorGML concept to an indoor patrol service. J. Sens. **2017**, 14 (2017)

7. Open Geospatial Consortium: OpenGIS Location Service (OpenLS) ImplementationSpecification: Core Services, OGC 07-074 (2008)

8. Open Geospatial Consortium: OGC Indoor Geographic Markup Language (IndoorGML), OGC 14–005r5 (2018)

9. Ozdikis, O., Orhan, F., Danismaz, F.: Ontology-based recommendation for points of interest retrieved from multiple data sources. In: Proceedings of the International Workshop on Semantic Web Information Management, p. 1. ACM (2011)

10. Park, J., Kang, H.Y., Lee, J.: A spatial-temporal POI data model for implementing location-based services. J. Korean Soc. Surv. Geodesy Photogram. Cartogr. **34**(6), 609–618 (2016)

11. Ruta, M., Scioscia, F., Ieva, S., De Filippis, D., Di Sciascio, E.: Indoor/outdoor mobile navigation via knowledge-based POI discovery in augmented reality. In: 2015 IEEE/WIC/ACM International Conference on Web Intelligence and Intelligent Agent Technology (WI-IAT), vol. 3, pp. 26–30. IEEE (2015)

12. Snowdon, C., Kray, C.: Exploring the use of landmarks for mobile navigation support in natural environments. In: Proceedings of the 11th International Conference on Human-Computer Interaction with Mobile Devices and Services, pp. 1–10. ACM (2009)
13. Trapp, M., Schneider, L., Holz, N., Döllner, J.: Strategies for visualizing points-of-interest of 3D virtual environments on mobile devices. In: Proceedings of the Sixth International Symposium on LBS & TeleCartography. Citeseer (2009)

CoolPath: An Application for Recommending Pedestrian Routes with Reduced Heatstroke Risk

Tianqi Xia[1,2(✉)], Adam Jatowt[2,3], Zhaonan Wang[1,2], Ruochen Si[1], Haoran Zhang[1], Xin Liu[2], Ryosuke Shibasaki[1], Xuan Song[1], and Kyoung-sook Kim[2]

[1] Center for Spatial Information Science, The University of Tokyo, Kashiwanoha, Kashiwa, Chiba 277-0882, Japan
xiatianqi@csis.u-tokyo.ac.jp

[2] National Institute of Advanced Industrial Science and Technology, Aomi, Koto, Tokyo 135-0064, Japan

[3] Kyoto University, Yoshida-honmachi, Sakyo-ku, Kyoto 606-8501, Japan

Abstract. Global warming and urbanization have made heatstroke a serious emergency disease especially for large cities in summer daytime. Although a lot of studies have focused on the heat-related analysis or on developing general routing applications for pedestrians, few have aimed at providing routing services for pedestrians specifically to reduce their heatstroke risk. In this research, we propose a novel routing system that can recommend pedestrian routes based on the estimated heatstroke risk using heterogeneous data and we conduct a detailed system design for the proposed application.

Keywords: Heatstroke · Google Street View · Routing system · Big data · Risk management

1 Introduction

With global warming and fast urbanization, the danger of heatstroke is getting increasingly serious all over the world, especially in large cities with urban heat island effects and high people flow. The previous studies on heatstroke have been mainly conducted from the perspective of macro-scope analysis with a top-down structure of the low-resolution census [7]. Though these studies can depict a long-term macro-scope trend of heatstroke risk, they pay less attention to analyzing heatstroke risk on the individual level and to providing solutions for reducing the heatstroke risk of individuals, among which outdoor pedestrians during summer should be important research targets since they have high chance to be exposed to solar radiation. In such a case, it is fundamental for outdoor pedestrians to choose a route with less heatstroke risk.

The study of individual heatstroke risk on road network level used to be difficult due to data limitation. However, with the development of location-based

© Springer Nature Switzerland AG 2020
S. Di Martino et al. (Eds.): W2GIS 2020, LNCS 12473, pp. 14–23, 2020.
https://doi.org/10.1007/978-3-030-60952-8_2

services (LBS), open data acquired from LBS applications with the higher spatial and temporal resolution have been applied to conduct road network level heat-related analysis [11,12] and have been applied in several routing systems concerning public health [18]. Russig and Bruns [16] propose a route planner to provide dynamic routes with minimal heat exposure and suggestions of outdoor activity time. However, their heatstroke risk assessment is based only on temperature index measured by weather stations, which, as a result, has low spatial resolution and neglects the landscape information.

In view of the above-mentioned concern, this study aims at proposing a novel pedestrian routing system named CoolPath that can recommend routes with reduced heatstroke risk. In this system, the network-based risk is calculated by the classical risk model with heterogeneous data sources that include Google Street View (GSV), OpenStreetMap road network and digital elevation model (DEM) data. The contributions of this study are as follows:

1. This study utilizes big heterogeneous data for analyzing a road-level heatstroke risk. The approach of combining large SVF data collection and risk modeling is novel for route recommendation.
2. A full-stack application prototype based on the system proposal is provided with the study area of Tokyo. In addition, the application has the function of setting customized study area, which makes it applicable to other regions.
3. Hiking function applied in the system makes it applicable to estimate heat-related risk in mountainous areas, which makes it possible for analyzing heatstroke risk in the rural areas of Japan.

The remainder of this paper is structured as follows: Sect. 2 introduces the methodology utilized in our routing service; Sect. 3 introduces data collection and preprocessing in the system. It also demonstrates the prototype and gives some examples on how the system works. Section 4 concludes the research and points out the limitations and future studies concerning this research.

2   Risk Model

The traditional emergency risk model is made up of three components: hazard, vulnerability, and exposure [9]. A simple approach for measuring emergency risk could be denoted by a multiplication of three factors: $R = r_{hazard} * r_{vulnerability} * r_{exposure}$. Generally, hazard represents the possibility that the emergency happens [10]; vulnerability represents the lack of resistance to emergency; exposure refers to the number of people or amount of time spent exposed to the hazard. The remainder of this section will respectively introduce these three parts in the context of heatstroke and finally introduce the proposed approach for applying risk model into a routing service.

2.1   Measuring Heatstroke Hazard

The hazard of heat-related risks could be measured by a comprehensive index involving several hazard factors. Wet Bulb Globe Temperature (WBGT) is currently the most widely used heat stress index which includes the factors of temperature, humidity, wind speed and radiation. The higher WBGT value indicates the higher possibility of heatstroke emergency [7].

Generally, in a citywide range WBGT does not have too many spatial differences due to the fact that a city shares similar spatial weather features and the number of stations measuring WBGT is limited. However, since the different conditions of temperature and radiation tend to occur at different hours of a day, a large distinction could be observed for different time spans within the same day, which indicates the importance of taking time into consideration. Therefore, in this study, we choose a data-driven approach to measure heatstroke hazard via historical WBGT data at different hours during different days. We utilize a normalized index P_t to represent the probability of time span that exceeds the warning temperature in all summer days.

2.2   Measuring Heatstroke Vulnerability

In the previous studies, heat vulnerability index (HVI) has been proposed for measuring the spatial vulnerability to heat-related disease [13] with several indicators mainly from the perspective of socio-economic and environment. Since our work focuses on recommending routes to individuals with the context of their current location, more attention should be paid to the environmental factors. Generally, environmental factors could be divided into the factors V_i that can increase risk, such as the environment that makes pedestrians be directly exposed to solar radiation, and the factors V_r that can reduce the risk such as water supply. In this study, we measure the vulnerability risk by the combination of sky view factor and accessibility model.

Sky view factor (SVF) is a widely utilized indicator for measuring urban environment and heat island effects. The previous studies have proved a strong relationship between SVF and urban temperature in different areas [1]. They mainly measure SVF using two data sources: the top-down approach applies image processing methods on satellite images [4]. In contrast, the bottom-up approach utilizes the data acquired from the surface including some surface models [8] and hemi-spheric photographs [5]. The spatial resolution of satellite images and surface models depends on the data source and is usually not suitable for road level SVF calculation. Hemi-spheric photographs are usually collected by driving throughout the streets. Thus this approach could ensure a high spatial resolution of road level SVF calculation. However, field survey for collecting hemi-spheric photographs in a metropolis like Tokyo is difficult due to the high cost. In recent years, Google Street View (GSV) has been proved to be an efficient data source for SVF calculation [11,12]. Compared to the field survey, GSV data could be collected through crawling images from the Internet, which makes it feasible to calculate SVF in a city-wide range. We then collect GSV data and process it for calculating SVF value. The detailed processing approach is introduced in the next section.

On the other hand, accessibility to POIs providing resting area and water supply could be calculated by a gravity model summing up the number of POIs with distance decay [6]. Thus in this study for each road segment s in time span t, vulnerability is calculated by e.q. $\frac{V_{ist}}{V_{rst}}$ where V_{ist} is denoted by the normalized SVF value, while V_{rst} is the normalized accessibility value during the time period.

2.3 Measuring Heatstroke Exposure

For pedestrians, exposure risk could be denoted by the time they are exposed to the high temperature during their walk. Thus given a route in outdoor space, exposure risk should be proportional to the road length divided by walking velocity. In the previous studies, walking speed is usually regarded as unchanged, and then walking time is proportional to the route distance [14]. However since risk measurement is sensitive to traveling time and the previous studies have pointed out the relationship between slope and route choices [15], we apply Tobler's Hiking function [17] to estimate pedestrian speed via slope information calculated by digital elevation model (DEM). Thus, the travel time t_s for one road segment s is computed by Eq. 1.

$$t_s = \frac{L_s}{6e^{-3.5|\frac{dh_s}{dx_s}+0.05|}} \tag{1}$$

where L_s is the length of road segment s, while dh_s and dx_s respectively denote the height and length difference in road segment s.

2.4 Risk Calculation

Using the risk factors introduced above, the total heatstroke risk R_{rt} for a candidate route r during a time span t is denoted as a sum of emergency risk value of each segment s with the control parameters a, b, c:

$$R_{rt} = \sum_{s \in r} (P_t)^a \left(\frac{V_{ist}}{V_{rst}}\right)^b (t_s)^c \tag{2}$$

Based on the equation for calculating the risk value, we utilize Dijkstra's algorithm to obtain the potential route candidates and then to recommend routes with the lowest values to users.

3 System Prototype: A Case Study of Tokyo

In this study, a prototype of CoolPath is developed for recommending routes in Tokyo city. In this section, we introduce the data and the preprocessing approach we utilized, and then show the user interface and the currently available functionalities.

3.1   Data Source and Preprocessing

OpenStreetView Data

POI data and Road network data have been collected from OpenSreetView. We utilize OSMnx [2] to collect road network data and Osmosis[1] in central Tokyo area, which includes 23 wards in the area of 619 km². We remove the high expressways and motorways which are not walkable finally obtaining around 756,760 road segments for our study area.

Temperature Data

Three years of hourly WBGT[2] and temperature data[2] are collected from the government websites. We mark the hours with WBGT higher than 25 °C and solar radiation higher than 0 as the hours that hazard happened.

Google Street View Data

The raw GSV data is represented as the panorama photographs with unique panorama ids. To collect GSV data, we firstly split the road network into five-meter intervals and extract the interpolated points for querying the id of target panorama photos. Then with the acquired id and meta data, we collect data from GSV websites and remove those panoramas which are far away from the points or in seasons outside of summer. We collect 1 million photos over all the Tokyo 23 wards that could be matched to OSM road network.

To calculate SVF from GSV data, we refer to the procedure applied by Li and Ratti [11] and Liang et al. [12]. At first we segment images using deep learning models to extract sky area. We test several deep learning models including FCN, SegNet, PSPNet and UNet and finally choose the SegNet model with the encoder of pretrained residual neural network (ResNet) as our image segmentation model, and we train the model based on car camera data from Cityscape Dataset [3].

Since the original GSV data is a panorama image taken from the special street view camera with an equirectangular projection which cannot preserve true area for calculating SVF, thus images should be converted to an equal-area projection. In particular, we transform the panorama images to the fisheye images with an azimuthal projection. Next, SVF is calculated based on the pixel numbers in the fisheye images. An example of GSV data collection and processing is visualized in Fig. 1.

Digital Elevation Data

Digital elevation data is utilized for computing velocity. Considering the data availability, in the current stage we use the SRTM dataset with a 30-m spatial resolution and match SRTM data to each road segment. In order to calculate slope from the SRTM dataset, we utilize a spline interpolation to smooth the result of SRTM in each road, and then a road segment is divided by several uphills and downhills. The Tobler's Hiking function is used in each slope to

---

[1] https://wiki.openstreetmap.org/wiki/Osmosis.

[2] https://www.wbgt.env.go.jp/record_data.php?region=03&prefecture=44&point=44132

[2] http://www.data.jma.go.jp/obd/stats/etrn/index.php

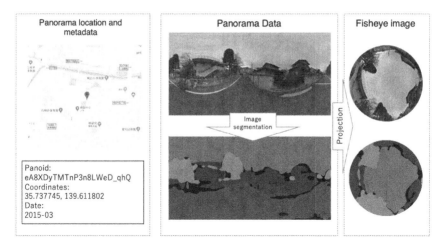

Fig. 1. An example of GSV data processing

estimate the traveling time over this slope. Finally, the traveling time for each road segment is generated by the sum of traveling time for each slope.

3.2   System Design and User Interface

With the data and methodology mentioned above, a web application based on the system overview shown in Fig. 2 is designed for providing route suggestions and risk visualization. As shown in the system overview, several functions concerning heatstroke risk analysis and routing service are provided in the prototype. They are listed in Table 1.

Based on the modules that were introduced above, a browser/client structure prototype has been developed with the front-end user interface shown in Fig. 3. In the current system, a Google-Map-style user interface is built with the components of a full-size map, a sidebar, and several map controls. The sidebar and controls are set to be foldable for a user to shift from the map view. The function views modules could be switched by clicking the tabs on the right side of the side bar.

All modules are made up of one or several components on the map with spatial visualization in map content and statistic visualization in controls and sidebars. The module visualization is illustrated in Fig. 4 and the current features of each module are listed as follows:

Route recommendation module generates the route with origin and destination coordinates as well as the query date type. The input coordinates are firstly snapped to nodes in road network and reverse-geocoded to the node address. The query results include the least heatstroke risk route on the map and the line chart in the figure control show the elevation or risk information in each road segments.

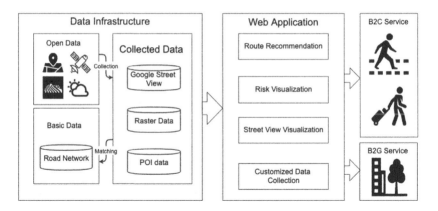

Fig. 2. CoolPath system overview

Table 1. Modules and functions

Module	Function	Objective
Route recommendation	Route query	Given origin, destination and time span, finding routes with less heatstroke risk
Risk visualization	WBGT query	Given date and time span, visualizing WBGT index values
	Road segment query	Visualizing the risk status of road segments
Street view visualization	GSV query	Given location information, providing metadata, processed images of panorama
Customized data acquisition	User-customized online query	Given query boundaries, providing the whole solution within the boundaries

GSV visualization module generates a panorama metadata list in side bar with query location as the input. GSV data can be downloaded and analyzed when selecting the row. The queried GSV point is visualized on the map with a pop-up showing the origin data, fisheye images and detected results.

Risk visualization module includes WBGT visualization and road segments visualization. As shown in Fig. 4, WBGT is visualized by line charts with different date types as the input. The overall road segments are visualized by vector tiles and the individual road segments could be queried by the coordinate input. WBGT line charts are shown in the figure control, and road segments are visualized as a layer on the map with attribute information in the sidebar and detailed value visualization given in the figure control.

Fig. 3. CoolPath system user interface: the main figure shows the map UI with folded controls and the sub figures shows the control contents.

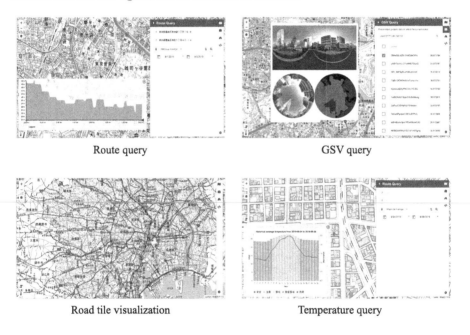

Route query   GSV query

Road tile visualization   Temperature query

Fig. 4. CoolPath system functions visualization

User-customized data acquisition module reads several types of boundary data and based on the used boundaries it collects the required data for CoolPath. The boundaries are visualized as a map layer and some of the results would be downloadable for local usage.

4   Conclusion

This study proposes a routing system for recommending routes to pedestrians that could minimize the total heatstroke risk. With the system proposed, this study designs an application for routing and visualizing the risk factors in Tokyo with heterogeneous data collected from different data sources.

Although this novel system is based on the combination and modification of several previous studies, it still has some limitations. First, we utilize the Cityscape Dataset for training image detecting model, as the street view data in the training dataset is the flat image which is different from the panoramas of GSV. This might cause the loss of accuracy for SVF calculation. In addition, due to the limitation of data and computing resources, we have not analyzed the temporal differences of vulnerability and exposure risk factors.

In the future, in addition to completing the prototype construction, we will try to apply more heterogeneous data to the routing system for more detailed risk evaluation and to build a smartphone application that can further benefit pedestrians in reducing heatstroke risk at real-time.

Acknowledgement. This work was supported by JSPS Grant-in-Aid for Early-Career Scientists (Grant Number 19K20352); the New Energy and Industrial Technology Development Organization (NEDO).

References

1. Ahmadi Venhari, A., Tenpierik, M., Taleghani, M.: The role of sky view factor and urban street greenery in human thermal comfort and heat stress in a desert climate. J. Arid Environ. **166**, 68–76 (2019). https://doi.org/10.1016/j.jaridenv.2019.04.009
2. Boeing, G.: OSMnx: new methods for acquiring, constructing, analyzing, and visualizing complex street networks. Comput. Environ. Urban Syst. **65**, 126–139 (2017). https://doi.org/10.1016/j.compenvurbsys.2017.05.004
3. Cordts, M., et al.: The cityscapes dataset for semantic urban scene understanding. In: Proceedings of the IEEE Conference on Computer Vision and Pattern Recognition, pp. 3213–3223 (2016)
4. Dirksen, M., Ronda, R.J., Theeuwes, N.E., Pagani, G.A.: Sky view factor calculations and its application in urban heat island studies. Urban Clim. 30 (2019). https://doi.org/10.1016/j.uclim.2019.100498
5. Grimmond, C., Potter, S., Zutter, H., Souch, C.: Rapid methods to estimate sky-view factors applied to urban areas. Int. J. Climatol. **21**(7), 903–913 (2001). https://doi.org/10.1002/joc.659. http://doi.wiley.com/10.1002/joc.659
6. Hansen, W.G.: How accessibility shapes land use. J. Am. Inst. Plan. **25**(2), 73–76 (1959)
7. Kasai, M., Okaze, T., Yamamoto, M., Mochida, A., Hanaoka, K.: Summer heatstroke risk prediction for Tokyo in the 2030s based on mesoscale simulations by WRF. J. Heat Isl. Inst. Int. **12**(2), 61–67 (2017). http://www.heat-http://www.heat-island.jp/webjournal/JGM8SpecialIssue/P-8kasai.pdf
8. Kastendeuch, P.P.: A method to estimate sky view factors from digital elevation models. Int. J. Climatol. **33**(6), 1574–1578 (2013)

9. Koks, E.E., Jongman, B., Husby, T.G., Botzen, W.J.: Combining hazard, exposure and social vulnerability to provide lessons for flood risk management. Environ. Sci. Policy **47**, 42–52 (2015). https://doi.org/10.1016/j.envsci.2014.10.013

10. Kron, W.: Keynote lecture: Flood risk = hazard exposure vulnerability. Technical Report (2002)

11. Li, X., Ratti, C.: Mapping the spatio-temporal distribution of solar radiation within street canyons of Boston using Google Street View panoramas and building height model (2018). https://doi.org/10.1016/j.landurbplan.2018.07.011

12. Liang, J., Gong, J., Zhang, J., Li, Y., Wu, D., Zhang, G.: GSV2SVF-an interactive GIS tool for sky, tree and building view factor estimation from street view photographs. Build. Environ. 168 (2019), (2020). https://doi.org/10.1016/j.buildenv.2019.106475

13. Nayak, S.G., et al.: Development of a heat vulnerability index for New York State. Public Health **161**, 127–137 (2018). https://doi.org/10.1016/j.puhe.2017.09.006

14. Novack, T., Wang, Z., Zipf, A.: A system for generating customized pleasant pedestrian routes based on OpenStreetMap data. Sensors **18**(11), 3794 (2018)

15. Pingel, T.J.: Modeling slope as a contributor to route selection in mountainous areas. Cartogr. Geogr. Inf. Sci. **37**(2), 137–148 (2010)

16. Rußig, J., Bruns, J.: Reducing individual heat stress through path planning.GI_Forum 1, pp. 327–340 (2017)

17. Tobler, W.: Three presentations on geographical analysis and modeling. NCGIA (1993)

18. Zhang, Y., Kawai, Y., Siriaraya, P., Jatowt, A.: Rehab-path: recommending alcohol and drug-free routes. In: Proceedings of International Conference on Information and Knowledge Management, pp. 2929–2932. Association for Computing Machinery (2019). https://doi.org/10.1145/3357384.3357843

What Do We Actually Need During Self-localization in an Augmented Environment?

Fan Yang[1](✉) , Zhixiang Fang[1,2](✉) , and Fangli Guan[1]

[1] State Key Laboratory of Information Engineering in Surveying, Mapping and Remote Sensing, Wuhan University, 129 Luoyu Road, Wuhan 430079, Hubei, People's Republic of China
{yang.f,zxfang}@whu.edu.cn
[2] Collaborative Innovation Center of Geospatial Technology, 129 LuoyuRoad, Wuhan 430079, People's Republic of China

Abstract. Self-localization is the behavior of pedestrians when they are disoriented in wayfinding. The widely used AR (augmented reality) navigation technology makes people understanding their location more intuitively. The signals that can be used for self-localization include landmarks, road names, direction guidance, etc., but these elements cannot all be added on a limited size screen. This paper using eye tracking technology to conduct an experiment, aims at exploring the relationship between different augmented information combinations and pedestrian self-localization efficiency and success rate in augmented scenes. Results show that direction guidance is the easiest information to understand, landmark is an important factor in increasing pedestrian safety, and road name has very low help for self-localization and can even increase people's cognitive burden.

Keywords: Self-localization · Augmented reality · Eye tracking · Navigation

1 Introduction

Self-localization is the behavior that trying to localize ourselves when we are disoriented and do not know where we are on the map (Meilinger et al. 2007). This usually happens at the beginning of navigation or at decision points. People come up with many ways to reduce the cognitive burden of navigation (Brügger et al. 2019; Koletsis et al. 2017). Augmented reality navigation is one of the most effective ways if there are no display errors caused by various reasons, such as GPS instability (Wen et al. 2013). The augmented reality information in navigation generally includes landmark, direction guidance, route instruction, etc. (May et al. 2003). How to evaluate the scientific of these information? What do we actually need during self-localization in an augmented environment?

To explore this issue, we used eye tracking technology to conduct the experiment. Eye tracking is a method for studying human gaze behavior, such as visual search effectiveness(Van Der Lans et al. 2008) and attention allocate mechanism (Park 2019). It is

© Springer Nature Switzerland AG 2020
S. Di Martino et al. (Eds.): W2GIS 2020, LNCS 12473, pp. 24–32, 2020.
https://doi.org/10.1007/978-3-030-60952-8_3

widely used in the research of pedestrian navigation behavior (Wang et al. 2019; Zhang et al. 2014). Ehinger et al. (2009) recorded eye movements and evaluating computational models that predicting human fixations. Kiefer et al. (2014) observed the visual matching process between environment and map during self-localization with real-world mobile eye tracking and found that route knowledge is more important than survey knowledge. Tanabe and Yoshioka (2020) compared gazing pattern on conventional smartphone navigation and the AR navigation and concluded that the lack of a bird's eye view of the route can cause users' uneasy feeling.

In this study, eye tracking technology is used to compare and analyze the relationship between different augmented information combinations and the efficiency and success rate of pedestrian self-localization, for judging its importance and providing a reference for the design of augmented reality navigation.

2 Method

2.1 Participants

The study group consisted of 6 males and 9 females aged between 21 and 27 (mean age = 23, SD = 1.51). The observers were students at Wuhan University, majoring in Geographic Information System. The study area is 11.2 km away from the school and they are unfamiliar with the scenes we selected. All of the students had normal or corrected-to-normal vision. The experimental procedure was carefully explained to the students before the experiment.

2.2 Materials

We selected real street scenes at 15 decision points, each of them were displayed in 5 ways of augmentation to simulate self-localization when using AR navigation (see Fig. 1 for example, (a) to (e) represent 5 ways of augmentation). Figure 1 (a) is none augmentation, only the bird's eye view of map is used to indicate route and confirm location; (b) is landmark augmentation which adds salient landmarks on the basis of (a); (c) is direction guidance augmentation which adds direction guidance on the basis of (a); (d) is road name augmentation which adds road name and route instruction described by road name on the basis of (a); (e) is combined augmentation which combines all the augmented information from (a) to (d). The picture in the lower right corner of Fig. 1 represents AOI (area of interest), which is used to detect gaze points in interest areas for subsequent analysis.

The subjects were asked to locate themselves according to the information on the screen. We use a 90 Hz screen eye tracking device (Tobii 4c) to detect the subject's gaze position. The distance between the subject's eyes and the screen is controlled between 50–60 cm. According to (Hassan et al. 2007), the critical points for efficient navigation were FOVs of 32.1°, 18.4° and 10.9° (diam.). This experiment was displayed on a 22.1-inch screen, and each scene image covered a viewing angle of 33.8° × 45°, thereby ensuring effective navigation.

(a) None augmentation

(b) Landmark augmentation

(c) Direction guidance augmentation

(d) Road name augmentation

(e) Combined augmentation

AOIs

Fig. 1. Examples of augmentation interface used in the experiment.

2.3 Experiment Design and Procedure

Participants are welcomed, introduced to the experiment, and asked to complete the Santa Barbara Sense of Direction scale (SBSODS), which is a self-reporting assessment that evaluates people's environmental spatial abilities (Hegarty 2002). Scores from the SBSOD scale can then be used to determine the spatial abilities of the subjects (from poor to good). In addition, subjects were calibrated for the eye tracking system. When the calibration meets the experimental requirements, start the preparation experiment so that the subjects are familiar with the experimental process. After they were able to operate the experimental system proficiently, the preparation experiment was interrupted and the formal experiment was performed.

The steps for each experiment are as follows: ① Click the scene switch button with the corresponding serial number in the upper left corner of the screen to start the experiment; ② Display the experimental scene in full screen and the 2D map of the scene area in the bottom right corner; ④The subject locates themselves and clicks the location on the 2D map; ⑤A pop-up dialog prompts that this task is complete, with a question attached: Which information did you use to determine your location? ⑥ The subject answers the above two questions; ⑦ Repeat the above steps.

Record the eye movement data of the subjects during the experiment and calculate the distribution of gaze points, as shown in Fig. 2.

Fig. 2. The distribution of gaze points.

A subject cannot meet the same scene during the experiment. In order to ensure that the 5 ways of augmentation of 15 scenes are tested evenly, the subjects are divided into 5 groups. The experimental arrangement of each subject is shown in Fig. 3.

	P1	P2	P3	P4	P5	P6	P7	P8	P9	P10	P11	P12	P13	P14	P15
1	/	/	/	//	//	//	///	///	///	////	////	////	/////	/////	/////
2	/	/	/	//	//	//	///	///	///	////	////	////	/////	/////	/////
3	/	/	/	//	//	//	///	///	///	////	////	////	/////	/////	/////
4	/////	/////	/////	/	/	/	//	//	//	///	///	///	////	////	////
5	/////	/////	/////	/	/	/	//	//	//	///	///	///	////	////	////
6	/////	/////	/////	/	/	/	//	//	//	///	///	///	////	////	////
7	////	////	////	/////	/////	/////	/	/	/	//	//	//	///	///	///
8	////	////	////	/////	/////	/////	/	/	/	//	//	//	///	///	///
9	////	////	////	/////	/////	/////	/	/	/	//	//	//	///	///	///
10	///	///	///	////	////	////	/////	/////	/////	/	/	/	//	//	//
11	///	///	///	////	////	////	/////	/////	/////	/	/	/	//	//	//
12	///	///	///	////	////	////	/////	/////	/////	/	/	/	//	//	//
13	//	//	//	///	///	///	////	////	////	/////	/////	/////	/	/	/
14	//	//	//	///	///	///	////	////	////	/////	/////	/////	/	/	/
15	//	//	//	///	///	///	////	////	////	/////	/////	/////	/	/	/

/ None augmentation   // Landmark augmentation   ///// Combined augmentation

/// Direction guidance augmentation   //// Road name augmentation

Fig. 3. Experimental arrangement. Each row represents a subject and each column represents a scene.

2.4   Measured Variables

Concerning subjects' behavior, we measured (a) positioning accuracy, (b) gaze duration and frequency on AOIs, (c) subjective feedback during the experiment: the information used during self-localization.

3   Results and Discussion

The subjects' gaze duration and times on different AOIs and augmented group were compared by analyzing the gaze data detected by the eye tracking system. The following experimental results were obtained.

(1)   Relationship between self-localization success rate and sense of direction scale: If the self-localization success rate of the subjects is calculated using the error within 20 m as the criterion, it can be found that the success rate of 15 subjects is related to SOD ($r = 0.5096$, $p = 0.0523 < 0.1$).

(2)   Gaze distribution and completion time for successful and unsuccessful tasks: The distribution of gaze points of successful and unsuccessful tasks on different AOIs is basically the same ($p < 0.01$), indicating that the amount of information obtained is the same. As shown in Fig. 4, the completion time of the successful task is lower than that of the unsuccessful ones, indicating that the success rate will not increase with the increase of the judging time. Only clearer clues can improve the self-localization success rate.

(3)   Gaze distribution and completion time for different ways of augmentation: Fig. 5 and 6 shows the completion time and number of gaze points for different ways of augmentation. Time spent in ascending order is road name augmentation, none augmentation, landmark augmentation, combined augmentation, direction guidance augmentation. The time cost of road name augmentation is significantly longer than that of none augmentation, and the number of gaze points of road name augmentation is the largest, i.e., the number of saccades is the most. Indicating that invalid augmentation may cause greater recognition pressure and reduce the task completion efficiency. The time cost and number of gaze points of direction guidance augmentation is the least, so it is the easiest and most effective way among these augmentations.

For different augmentations, the distribution of the number of gaze points on each AOIs are similar, as shown in Fig. 7. In terms of the most efficient method, direction guidance augmentation, the gaze point on the bird's eye view is the most, because the location needs to be determined on it, followed by the landmark, which are far greater than the attention on the direction indication. Tanabe and Yoshioka (2020) conducted experiments using AR navigation with direction guidance and they found that people would felt uneasy without a bird's-eye view of the route. The interface area of navigation devices (smartphones) is limited, which is not enough to place a detailed bird's view of the route. Matching environmental landmarks with augmented landmarks can play a similar role, which is why the use of landmarks is significantly higher.

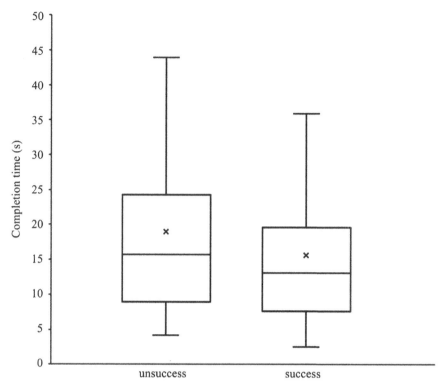

Fig. 4. Completion time for successful and unsuccessful tasks.

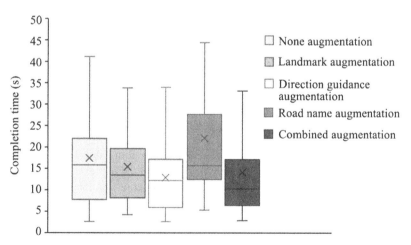

Fig. 5. Completion time for different ways of augmentation

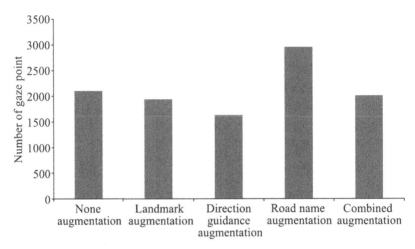

Fig. 6. Gaze distribution for different ways of augmentation

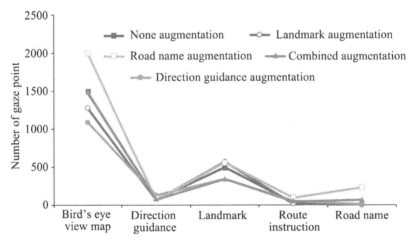

Fig. 7. Gaze distribution in AOIs for different ways of augmentation

(4) Subjective feedback during the experiment: As shown in Table 1, the use of land-marks is frequently regardless of the type of augmented expression, they are good references for self-localization. When giving direction guidance augmentation, pedestrians can quickly locate themselves without a landmark as a reference, result-ing in a reduction in the use of landmarks. Road name augmentation have little help in self-localization; it was only used once (among 90 tasks) when there were obvi-ous street signs in the field of vision. The shape of the road can be well combined with road decision point. Pedestrians can quickly determine their area by matching the shape of the road on the map, but the direction needs to be further judged.

Table 1. Subjective feedback about what information used for self-localization.

	None augmentation	Landmark augmentation	Direction guidance augmentation	Road name augmentation	Combined augmentation
Landmark	91.1%	97.8%	77.8%	88.9%	88.9%
Direction	0	0	57.8%	0	46.7%
Road name	0	0	0	2.2%	0
Road shape	17.8%	20%	4.4%	8.9%	8.9%

4 Summary and Conclusion

Several significant findings emerged from the present study. First, the self-localization success rate and completion efficiency are positively related to people's sense of direction, and the success rate can only be improved through clearer information provided. Second, direction guidance is the most intuitive augmentation, which can significantly improve the completion efficiency. Landmark is necessary, it can effectively increase the sense of security of pedestrians, and is a very suitable reference for self-localization. The bird's eye view of the route can also increase the sense of security of pedestrians, but the occlusion problem needs to be considered on the screen of a limited navigation device (Tanabe and Yoshioka 2020). In this experiment, the augmentation of road name makes no sense, and it even reduce the recognition efficiency.

In summary, direction guidance and landmark are the two most important pieces of information, which can basically meet the needs of self-localization; the bird's-eye view of the route can be represented by general direction; the road name can be ignored. We are planning experiments in other complex navigation scenarios, such as underground area and pedestrian overpass, to explore the importance of augmentation during AR navigation to avoid cognitive pressure caused by ineffective augmentation.

Acknowledgement. The research was supported in part by the National Natural Science Foundation of China (Grants 41771473, 41231171) and National key R&D plan (Grant 2017YFC1405302).

References

Brügger, A., Richter, K.-F., Fabrikant, S.I.: How does navigation system behavior influence human behavior? Cogn. Res. Principles Implications **4**(1), 1–22 (2019). https://doi.org/10.1186/s41235-019-0156-5

Ehinger, K.A., Hidalgo-Sotelo, B., Torralba, A., Oliva, A.: Modelling search for people in 900 scenes: a combined source model of eye guidance. Visual Cogn. **17**, 945–978 (2009). https://doi.org/10.1080/13506280902834720

Hassan, S.E., Hicks, J.C., Lei, H., Turano, K.A.: What is the minimum field of view required for efficient navigation? Vision. Res. **47**, 2115–2123 (2007). https://doi.org/10.1016/j.visres.2007.03.012

Hegarty, M.: Development of a self-report measure of environmental spatial ability. Intelligence. **30**, 425–447 (2002). https://doi.org/10.1016/S0160-2896(02)00116-2

Kiefer, P., Giannopoulos, I., Raubal, M.: Where am I? investigating map matching during self-localization with mobile eye tracking in an urban environment. Trans. GIS **18**, 660–686 (2014). https://doi.org/10.1111/tgis.12067

Koletsis, E., van Elzakker, C.P.J.M., Kraak, M.-J., Cartwright, W., Arrowsmith, C., Field, K.: An investigation into challenges experienced when route planning, navi-gating and wayfinding. Int. J. Cartography **3**, 4–18 (2017). https://doi.org/10.1080/23729333.2017.1300996

May, A.J., Ross, T., Bayer, S.H., Tarkiainen, M.J.: Pedestrian navigation aids: infor-mation require-ments and design implications. Pers. Ubiquit. Comput. **7**, 331–338 (2003). https://doi.org/10.1007/s00779-003-0248-5

Meilinger, T., Hölscher, C., Büchner, S.J., Brösamle, M.: How Much Information Do You Need? Schematic Maps in Wayfinding and Self Localisation. In: Barkowsky, T., Knauff, M., Ligozat, G., Montello, D.R. (eds.) Spatial Cognition 2006. LNCS (LNAI), vol. 4387, pp. 381–400. Springer, Heidelberg (2007). https://doi.org/10.1007/978-3-540-75666-8_22

Park, S.: Attention scales with object size. Nature Human Behav. **3**, 12–13 (2019). https://doi.org/10.1038/s41562-018-0497-y

Tanabe, A., Yoshioka, Y.: Gazing Pattern While Using AR Route-Navigation on Smartphone. In: Ahram, T. (ed.) AHFE 2019. AISC, vol. 973, pp. 325–331. Springer, Cham (2020). https://doi.org/10.1007/978-3-030-20476-1_33

Van Der Lans, R., Pieters, R., Wedel, M.: Eye-movement analysis of search effectiveness. J. Am. Stat. Assoc. **103**, 452–461 (2008). https://doi.org/10.1198/016214507000000437

Wang, C., Chen, Y., Zheng, S., Liao, H.: Gender and age differences in using indoor maps for wayfinding in real environments. ISPRS Int. Geo-Inf. **8**, 11 (2019). https://doi.org/10.3390/ijgi8010011

Wen, J., Helton, W.S., Billinghurst, M.: A study of user perception, interface performance, and actual usage of mobile pedestrian navigation aides. Proc. Hum. Factors Ergon. Soc. Annu. Meet. **57**(1), 1958–1962 (2013). https://doi.org/10.1177/1541931213571437

Zhang, X., Li, Q., Fang, Z., Lu, S., Shaw, S.: An assessment method for landmark recognition time in real scenes. J. Environ. Psychol. **40**, 206–217 (2014). https://doi.org/10.1016/j.jenvp.2014.06.008

Predicting Indoor Location based on a Hybrid Markov-LSTM Model

Peixiao Wang[1]([✉]) [iD], Sheng Wu[1] [iD], and Hengcai Zhang[2]([✉]) [iD]

[1] The Academy of Digital China, Fuzhou University, Fuzhou 350002, China
peixiao_wang@163.com
[2] State Key Lab of Resources and Environmental Information System, Institute of Geographical Sciences and Natural Resources Research, Chinese Academy of Sciences, Beijing 100101, China
zhanghc@lreis.ac.cn

Abstract. To overcome the problem of dimension curse in the processing of predicting indoor location by using the traditional Markov chains, this paper proposes a novel hybrid Markov-LSTM model to predict the indoor user's next location, which adopt the multi-order Markov chains (k-MCs) to model the long indoor location sequences and use LSTM to reduce dimension through combining multiple first-order MCs. Finally, we conduct comprehensive experiments on the real indoor trajectories to evaluate our proposed model. The results show that the Markov-LSTM model significantly outperforms five existing baseline methods in terms of its predictive performance.

Keywords: Indoor location prediction · Movement trajectory · Markov-LSTM

1 Introduction

As a classical statistical model, the first-order Markov chain (1-MC) has strong interpretability and is widely used in location prediction. However, 1-MC assumes that the location at the next moment is only related to the current location, which significantly limits the predictive performance of the model [1, 2]. To address this deficiency, the multi-order Markov chain (k-MC) is proposed [3]. The k-MC assumes that the location at the next moment is related to the previous k locations but is prone to problems with dimensionality disaster, i.e., its state space explodes with an increase in k, which renders k-MC less practical in the field of location predictions [4].

Therefore, we propose a hybrid Markov-LSTM model. The model study attempts to combine the advantages of the Markov and LSTM models to improve the performance of the location prediction model. This study makes several significant contributions, which are summarized as follows:

(1) A new multi-step Markov transition probability matrix, which divides the multi-order Markov model into multiple first-order models and solves the shortcomings of the multi-order Markov model in the dimension disaster.
(2) Fusion of the prediction results of the multiple first-order Markov models based on the advantages of the LSTM for predicting long-sequence data. This improved the practicality of the multi-order Markov model for location prediction.

© Springer Nature Switzerland AG 2020
S. Di Martino et al. (Eds.): W2GIS 2020, LNCS 12473, pp. 33–38, 2020.
https://doi.org/10.1007/978-3-030-60952-8_4

2 Methodology

The structure of Markov-LSTM is presented in Fig. 1. Our method is divided into four phases: location sequence detection, multi-step transition probability matrix definition, adjacent locations selection, and fusion multiple Markov chains.

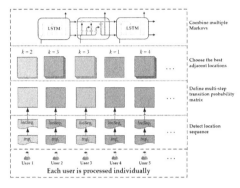

Fig. 1. Markov-LSTM model overall process.

2.1 Definition of the multi-step transition probability matrix

In this study, we used the indoor-STDBSCAN [5] and the nearest-neighbor search to convert the trajectory into a location sequence. In order to improve the practicability of k-MC in location prediction, we propose a novel k-step Markov chain, $MC^{(k)}$.

Definition 1 (1-Step Transition Probability Matrix). The 1-step transition probability matrix, $Y^{u(1)}$, of user u is equivalent to the 1-order transition probability matrix, $Y_{ij}^{u(1)}$, which represents the probability that user u moves from location l_i through one step to location l_j. $Y_{ij}^{u(1)}$ can be defined with the following expression:

$$Y_{ij}^{u(1)} = \frac{\sum_{p=1}^{m-1}\left|\left\{l_p^u=l_i\cap l_{p+1}^u=l_j\right\}\right|}{\sum_{p=1}^{m}|\{l_p^u=l_i\}|} \quad \begin{array}{l} Y^{u(1)} \in \mathbb{R}^{N\times N} \\ l_p^u \in locSeq^u \end{array} \tag{1}$$

where $locSeq^u$ represents the location sequence, $\{l_i^u\}_{i=1}^m$, of user u, $\sum_{p=1}^{m-1}\left|\left\{l_p^u = l_i \cap l_{p+1}^u = l_j\right\}\right|$ represents the distance that user u moves from location l_i through one step to location l_j, $\sum_{p=1}^{m}|\{l_p^u = l_i\}|$ represents the total distance that user u moves from location l_i through one step to other locations, and N represents the total number of shops in the mall.

Definition 2 (k-Step Transition Probability Matrix). The k-step transition probability matrix, $Y^{u(k)}$, of user u is a $N \times N$ matrix, $\hat{y}^{u(l_i \rightarrow *:k)} = Y_{i \rightarrow *}^{u(k)}$, which represents the probability that user u moves from location l_i through k steps to other locations. The definitions of $Y^{u(k)}$ and $\hat{y}^{u(l_i \rightarrow *:k)}$ for user u can be expressed with Eqs. (2) and (3), respectively:

$$Y^{u(k)} = P(L^u_{m+1}|L^u_{m-k+1}) \, Y^{u(k)} \in \mathbb{R}^{N \times N} \tag{2}$$

$$\hat{y}^{u(l_i \to *:k)} = P(L^u_{m+1}|L^u_{m-k+1} = l^u_{m-k+1}) \, \hat{y}^{u(l_i \to *:k)} \in \mathbb{R}^{1 \times N} \tag{3}$$

where $Y^{u(k)}$ can be directly obtained by $Y^{u(1)}$, i.e. $Y^{u(k)} = \left(Y^{u(1)}\right)^k$, L^u_{m-k+1} represents a random variable of user u, $L^u_{m-k+1} = l^u_{m-k+1}$ indicates that user u determines to visit location l^u_{m-k+1} at random variable L^u_{m-k+1} (l^u_{m-k+1} can be obtained in the location sequence $locSeq^u$), $Y^{u(k)}$ describes the effect that cross-location has on the prediction results from another perspective.

2.2 Selection of the best adjacency locations

Similar to the k-MC, the Markov-LSTM model must also determine the hyperparameter, k, i.e., the number of locations that the prediction result depends on. This value is usually determined using cross-validation to minimize the model prediction error [6, 7]. Taking user u with a k value of k_u as an example, when $k_u > 1$, the k-MC can be decomposed based on the following expressions.

$$\begin{cases} \hat{y}^{u(l^u_m \to *:1)} = P\left(L^u_{m+1}|L^u_m = l^u_m\right) \\ \hat{y}^{u(l^u_{m-1} \to *:2)} = P\left(L^u_{m+1}|L^u_{m-1} = l^u_{m-1}\right) \\ \qquad \qquad . \\ \qquad \qquad . \\ \qquad \qquad . \\ \hat{y}^{u\left(l^u_{m-k_u+1} \to *:k_u\right)} = P\left(L^u_{m+1}|L^u_{m-k_u+1} = l^u_{m-k_u+1}\right) \end{cases} \tag{4}$$

where $\left\{\hat{y}^{u\left(l^u_{m-i+1} \to *:i\right)}\right\}_{i=1}^{k_u}$ represents the prediction results of multiple first-order Markov models for user u.

2.3 Fuse multiple Markov models

For each user, u, we have established k_u first-order Markov models. Therefore, this study combines k_u first-order Markov models to ensure location prediction performance. Considering the order of the k_u first-order Markov model prediction results, i.e. $\left\{\hat{y}^{u\left(l^u_{m-i+1} \to *:i\right)}\right\}_{i=1}^{k}$, we use the LSTM model to fuse k_u results.

3 Experimental results and analysis

3.1 Data sources

The experimental data consisted mainly of Wi-Fi positioning data for 50 users and shop data for a shopping mall in Jinan City, China. The data covered the eight floors of the shopping mall from December 20, 2017, to February 1, 2018. Data for each trajectory included the unique identifier of the user, record upload time, the user's (X, Y) coordinates, and the unique floor identifier. There are 489 shops in the mall. Data for each shop included the shop's unique ID, the shape of the shop (a polygon composed of the coordinate sequence), the shop name, and the floor ID.

3.2   Evaluation metrics and comparative methods

In this study, we treats location prediction as a classification problem. Using *Accuracy@X*, *Precision@X*, *Recall@X*, and *F1 − Measure@X* (top *X* locations) as quantitative indicators of the evaluation model [8].

To comprehensively evaluate the performance of the Markov-LSTM model, we used five baseline methods for comparison: 1-MC, HMM (Hidden Markov model), RNN (Recurrent neural network), LSTM (Long-short-term-memory network), and GRU (Gated-recurrent-unit network). The prediction performance of HMM, RNN, LSTM, and GRU is related to the number of hidden states. In the comparison experiment, the number of states in HMM was varied among 10, 15, and 20 states. The number of hidden states in RNN, LSTM, and GRU was varied among 64, 128, and 256 states.

3.3   Comparison with baselines

In this section, Fig. 2 compares the prediction performance of the five models.

(1) From an overall perspective. If we take $X = 3$ as an example, the average *Accuracy@3*, *Precision@3*, *Recall@3*, and *F1 − Measure@3* of the 1-MC and HMM were 39.64%, 36.71%, 35.21%, and 35.87%, respectively. The average *Accuracy@3*, *Precision@3*, *Recall@3*, and *F1 − Measure@3* of the RNN, LSTM, and GRU were 64.74%, 62.10%, 55.91%, and 58.84%, respectively. Compared with the Markov-LSTM, the four indicators for the Markov-LSTM improved by 7.33%, 7.47%, 5.46%, and 6.38%, respectively.

(2) From a local perspective, the 1-MC model achieved poor prediction performance, with *Accuracy@3*, *Precision@3*, *Recall@3*, and *F1 − Measure@3* at 28.64%, 24.77%, 26.36%, and 25.54%, respectively. The LSTM model achieved good predictive performance, with *Accuracy@3*, *Precision@3*, *Recall@3*, and *F1 − Measure@3* at 67.79%, 65.78%, 55.15%, and 55.99%, respectively. Overall, the Markov-LSTM model improved indoor location prediction performance significantly by enhancing the *Accuracy@3* by between 6.29 and 43.43%, *Precision@3* by between 3.79 and 44.8%, *Recall@3* by between 9.23 and 35.02%, and the *F1 − Measure@3* by between 13.8 and 39.68%.

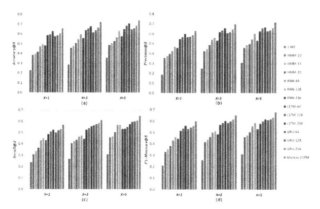

Fig. 2. Comparisons of the baselines using the dataset: (a) location prediction accuracy, (b) location prediction precision, (b) location prediction recall, and (d) location prediction f1-measure.

4 Conclusions

In this study, we proposed a novel hybrid Markov-LSTM model for indoor location prediction. During experiments, we conducted a comparison with five existing baseline methods, including the MC, HMM, RNN, LSTM, and GRU models. Compared with the existing methods, the Markov-LSTM model significantly improved indoor location prediction performance by enhancing the $Accuracy@3$ by between 6.29 and 43.43%, $Precision@3$ by between 3.79 and 44.8%, $Recall@3$ by between 9.23 and 35.02%, and the $F1 - Measure@3$ by between 13.8 and 39.68%. This demonstrates the efficiency of the Markov-LSTM model.

Funding. This project was supported by National Key Research and Development Program of China, (Grant Nos. 2016YFB0502104, 2017YFB0503500), and Digital Fujian Program (Grant No. 2016-23).

References

1. Gambs, S., Killijian, M.-O., Nunez del Prado Cortez, M.: Next place prediction using mobility markov chains. In: Proceedings of the 1st Workshop on Measurement, Privacy, and Mobility, MPM 2012, ACM (2012) https://doi.org/10.1145/2181196.2181199
2. Gambs, S., Killijian, M.O., del Prado Cortez, M.N.: Show me how you move and i will tell you who you are. Trans. Data Privacy **4**(2), 103–126 (2011)
3. Sha, W., Zhu, Y., Chen, M., Huang, T.: Statistical learning for anomaly detection in cloud server systems: a multi-order markov chain framework. IEEE Trans. Cloud Comput. **6**(2), 401–413 (2015)
4. Yu, X.G., Liu, Y.H., Da, W., Lei, L.Y.: A hybrid markov model based on EM algorithm. In: International Conference on Control (2006)
5. Peixiao, W., Sheng, W., Hengcai, Z., Feng, L.: Indoor location prediction method for shopping malls based on location sequence similarity. ISPRS Int. J. Geo-Inf. **8**(11), 517 (2019). https://doi.org/10.3390/ijgi8110517

6. Cheng, S., Lu, F., Peng, P., Wu, S.: Short-term traffic forecasting: an adaptive ST-KNN model that considers spatial heterogeneity. Comput. Environ. Urban Syst. pp. S0198971518300140 (2018)
7. Xia, D., Wang, B., Li, H., Li, Y., Zhang, Z.: A distributed spatial–temporal weighted model on MapReduce for short-term traffic flow forecasting. Neurocomputing **179**, 246–263 (2016)
8. Yang, Y.: An evaluation of statistical approaches to text categorization. Inf. Retrieval **1**(1–2), 69–90 (1999)

GeoSensor Data, Smart Environments, IoT, and Ambient Spatial Intelligence

Massive Spatio-Temporal Mobility Data: An Empirical Experience on Data Management Techniques

Sergio Di Martino$^{(\boxtimes)}$ and Vincenzo Norman Vitale

University of Naples "Federico II", DIETI, 80127 Naples, Italy
{sergio.dimartino,vincenzonorman.vitale}@unina.it

Abstract. The technological improvements within the Intelligent Transportation Systems, based on advanced Information and Communication Technologies (like Smartphones, GPS handhelds, etc.), has led to a significant increase in the availability of datasets representing mobility phenomena, with high spatial and temporal resolution. Especially in the urban scenario, these datasets can enable the development of "Smart Cities". Nevertheless, these massive amounts of data may result challenging to handle, putting in crisis traditional Spatial Database Management Systems. In this paper we report on some experiments we performed to handle a massive dataset of about seven years of parking availability data, collected from the municipality of Melbourne (AU), being about 40 GB. In particular, we describe the results of an empirical comparison of the retrieval performances offered by three different off-the-shelf settings to manage these data, namely a combination of *PostgreSQL + PostGIS* with standard indexing, a clustered setup of *PostgreSQL + PostGIS*, and a combination of *PostgreSQL + PostGIS + Timescale*, a storage extension specialized in handling temporal data. Results show that the standard indexing is by far outperformed by the two other solutions, which anyhow have different trade-offs. Thanks to this experience, other researchers facing the problems of handing these kinds of massive mobility dataset might be facilitated in their task.

Keywords: Mobility datasets · Spatio-Temporal databases · Database indexing · Knowledge discovery

1 Introduction

According to some researches, by 2050, up to 70% of the world's population (more than 6 billion people) are expected to live in cities and surrounding regions [1, 2]. As a consequence, to handle such a massive amount of people, these growing and expanding cities must implement smart and innovative solutions to deliver services, increase operational efficiencies, and reducing management cost. Although there is not yet a formal and widely accepted definition of "Smart City", the common goal is to leverage on Information and Communication Technologies (ICTs) to make the city services and monitoring more

© Springer Nature Switzerland AG 2020
S. Di Martino et al. (Eds.): W2GIS 2020, LNCS 12473, pp. 41–54, 2020.
https://doi.org/10.1007/978-3-030-60952-8_5

aware, interactive, and efficient [3]. A widely accepted vision is that the *Internet of Things* (IoT) paradigm, the seamless on-line integration of a everyday devices anD sensors, will be a key driver to implement the concept of smartness of a city [4]. Indeed, the integration of IoT sensors into the urban infrastructure (transport, health, environment, etc.) could lead to a seamless urban environment monitoring, by harvesting information from the environments, to provide a range of novel services for citizens, and knowledge for policy makers [1]. As an example, *Data-driven* Intelligent Transportation Systems (ITS) are meant to leverage massive amounts of mobility data that are collected daily, to exploit the existing road infrastructure in a smarter way [5,6], or to offer new use cases, with the goal to improve the quality of life and safety for citizens [7].

Nevertheless, this fascinating scenario is posing significant challenges to the ICT community. Each sensor deployed in a smart city produces potentially useful geo-referred data, which needs to be transferred to a data center, stored, and subsequently processed, to be used for administrative analysis and optimization tasks [8]. The amount of harvested geo-referred data could reach unprecedented rates, as some studies foresee a total of 1 trillion interconnected sensors by 2030, and traditional data storage solutions may be in troubles in handle such amount of data [9,10]. To make things worse, to date we are far from having any unifying information management platform for smart cities, even if a lot of research efforts are devoted in that direction (e.g.: [1]). Rather, we are still at an artisan state, where each collected dataset require an ad-hoc technological pipeline to be pre-processed, in order to be suitable for further knowledge extraction.

Within a project aimed at improving parking availability predictions [5], in this paper we report on an empirical study we performed to assess some off-the-shelf solutions to store a massive spatio-temporal dataset containing seven years of parking availability data, collected from the municipality of Melbourne (AU). This is a typical kind of dataset for a smart city, collected by a thousands of sensors embedded in the environment, accounting for about 40 GB of raw data. The goal of the project is to develop a tool to interactively explore massive amounts of parking data, to support a Decision Maker in acquiring a deeper insight on the mobility phenomena [7], since data visualization techniques are widely recognized as powerful and effective in this domain [11,12], by exploiting human abilities to perceive visual patterns and to interpret them [13]. By storing this data in a widely used setting, composed of *PostgreSQL* and *PostGIS*, even in presence of meaningful indexes and optimizations, each spatio-temporal query took up to 10 min. As a consequence, it was impossible for us to develop such an interactive tools. This motivated us to empirically assess if the exploitation of an off-the-shelf *Time-Series Management System* [14] could bring significant benefits, wrt. standard Spatial Database Management Systems, as found in other domains (e.g.: [15]). Thus, using the Melbourne mobility dataset, we compared three data management settings for spatio-temporal datasets, aimed at supporting subsequent data analytics tasks. Two are based on the well known combination of *PostgreSQL* and *PostGIS*, and differ in the way data are physically ordered in the files. The last setting is composed by *PostgreSQL*, *PostGIS* and *Timescale*, so on top of the previous stack we added a Time-Series Management System [14], to explicitly handle the temporal dimension of the data. These three settings have been benchmarked to assess the achievable performances in typical retrieval tasks to extract new knowledge from this kind of datasets (e.g.: [12,16]).

The rest of the paper is organized as follows. In Sect. 2 we describe in details we spatio-temporal dataset we employed in our experiments. In Sect. 3 we describe the different settings, while in Sect. 4 we detail the experimental setup and the results. Finally Sect. 5 concludes the article discussing also future work.

2   Mobility Datasets

The sensor networks of Smart cities can generate amounts of spatio-temporal data at an unprecedented rate, and novel software architectures are required to handle such amount of data (e.g.: [1,2,17]). Indeed, it is not uncommon to have datasets of many Gigabytes per month. In these datasets, the temporal part represents a challenge for Data Management solutions [14], especially in combination with the spatial dimension. To deal with them, a great number of modern general purpose storage systems (e.g.Redis, HBase, MongoDB, Neo4j), designed to handle Big Data, have been tested in the field of spatio-temporal data handling [18]. On the other hand, consolidated solutions for spatial data management, namely PostgreSQL[19] and PostGIS[20], could be extended in order to face with challenges deriving from the massive aspect of temporal data from IoT, with capabilities offered by modern TSMS (i.e.Timescale [21]).

As an example of these massive datasets, in a past project on parking availability predictions, we exploited data collected from sensors installed by the municipality of San Francisco (USA), on about 8,000 on-street parking stalls [5]. Those sensors generated about 2 GB per month. Let us note that this was a pilot project [22], as the total number of parking stalls in San Francisco is more than 250,000. Thus, in a futuristic scenario where each stall is instrumented by a sensors, the amount of generated data would be in the range of 60 GB per month. The interested reader can find further details on the dataset, plus instructions on how to download a sample of data in [23].

The Municipality of Melbourne, the capital city of the Australian state of Victoria, deployed a great number of sensors for monitoring on-street parking availability. More than 4000 sensors, around 480 street segments across the Melbourne Central Business District (CBD) were installed, and the collected data were made publicly available, both as real-time data through some APIs, both as aggregate historical data, in downloadable format [24]. In Fig. 1 we report the distribution of sensors over the urban area of Melbourne CBD. Again, let us note that, like for San Francisco, also in this case the instrumented area is just a fraction of the total urban area.

We focused on a dataset containing such parking data, from 2011 to 2017, for a total of about 275 Million records. The conceptual model, expressed as an Entity Relationship Diagram, is reported in Fig. 2. More in detail, the database

Fig. 1. Melbourne on-street parking sensor map. From: Victoria State Government, City of Melbourne [24]

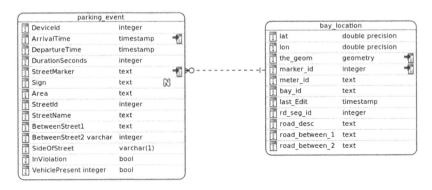

Fig. 2. Entity relation diagram for the dataset

schema is made up of two tables: *bay_location* and *parking_events*. The first one models the single parking stalls, thus containing information about the geographical location of the parking bay, its geometry, the road segment it belongs to, the shape, and the ID of the sensor. The second entity models the information about the parking availability, as monitored from the sensors. Here, each tuple describes the state of the parking spot, namely if it is free or occupied, plus some additional information to link it to a sensor and a parking bay. This table is mostly full of temporal data and represent the most prominent part of the dataset.

As a consequence, each spatio-temporal query performed on this dataset (e.g.: retrieve all the parking events in a range of 500 m from the Stadium on a specific day) requires to join these two tables, to get spatial filtering from *bay_locations* and temporal filtering and parking data from the *parking_events*.

When implemented in *PostgreSQL* with the spatial extension *PostGIS*, this dataset required about 40 GB of storage space, without considering the size for the indexing.

3   The Investigated Data Management Technologies

In this work we empirically assess the benefits and trade-offs of different settings to manage massive mobility data, including novel DataBase Management Systems oriented to Time-Series. The main requirement we had in our parking prediction project was to employ off-the-shelf data management solutions, thus discarding research products, which may lack of technical support. Let us note that, in our case, the sensors have fixed positions (corresponding to those of the parking stalls), so the outcome of the huge literature on spatio-temporal databases for Moving Objects is not adequate to this specific category of mobility dataset [25]. We also had to discard native Time-Series Management Systems (TSMS), like for instance the well-known *InfluxDB* [26] that we benchmarked in another paper [15], as these DBMS does not support for spatial primitives. As a consequence, in the end we focused our investigation on PostgreSQL, and its combination with spatial and temporal extensions, namely PostGIS and Timescale. In the following we describe the main characteristics of the three involved solutions.

PostgreSQL. PostgreSQL [19] is an Object Relational DBMS (ORDBMS), sparkling from a project of the University of Berkley, currently open source and multiplatform. To date, it is widely used in many different contexts and applications, for industrial, government and academic purposes (e.g.: [15,27,28]).

According to Brewer's *CAP Theorem* [29], PostgreSQL is classified as **AC**, since it provides high Availability and strong Consistency. Being based on the Relational data model, PostgreSQL supports a large part of the ISO/IEC SQL standard, as stated in the official on-line documentation, offering also a total compliance with ACID properties. As stated in official documentation *"[...] on the version 12 release in October 2019, PostgreSQL conforms to at least 160 of the 179 mandatory features for SQL:2016 Core conformance. As of this writing, no relational database meets full conformance with this standard"*.

A key feature of PostgreSQL is the possibility to add plug-ins to increment the capabilities and features offered by the DBMS. To date, a number of plug-ins are available, for many different domains, like spatial (*PostGIS*), temporal (*TimeScale*), Full Text Search engine (*OpenFTS*), etc.

PostGIS. PostGIS [20] is one of the key extensions of PostgreSQL, intended to add support for geographic objects and location queries [30] in a manner that is compliant with the *Simple Feature Access* specifications from ISO/Open Geospatial Consortium (OGC) [31]. Also PostGIS, like PostgreSQL, is open-source. PostGIS adds to PostGres a wide range of object expression methods, for example, use WKT (Well Known Text) and (Well Known Binary) to express different geometry types. In 2D coordinates, PostGIS supports 3D spatial data objects of WKT of OpenGIS to allow 3DZ, 3DM and 4D coordinates. As for data indexing, PostGIS adds support for *R-tree* spatial indexes [32]. Indeed, let us note that traditional index structures, based on the one-dimensional ordering of key-values, like the B-trees, do not work with spatial data, as the search space is multi-dimensional. R-Trees can be seen as a multi-dimensional extension of B-Trees, as they are a suitable structure to index n-dimensional objects, by approximating each (spatial) object by its bounding rectangle [33]. Like Post-Gres for B-Trees uses the implementation from [34], PostGIS uses a *Generalized Search Tree* (GiST) [35] based implementation of the indexes. GiST is a balanced, tree-structured access method, that acts as a base template in which to implement arbitrary indexing schemes, including R-Tree for GIS data used in PostGIS.

Timescale. Timescale [21,36] is a time-series database [14], intended as an extension of PostgreSQL. Being based on a relational model, Timescale transactions are compliant with ACID properties. Moreover it offers the advantages of being SQL-compliant to manage Time-Series data, along with the possibility to perform joins among Time Series and standard relational tables. Timescale physically orders records by time. The data model is mainly based on the classic relational one. Indeed values are stored in *Hyper Tables*, namely tables internally partitioned in smaller tables called *chunks*.

An Hypertable is partitioned horizontally on the basis of time, a maximum time period for each chunk will be established when the Hypertable is created, namely the chunk_time_interval. Then the (optional) vertical partitioning is performed on the basis of a meaningful column, by specifying the number of partitions or a partition function. In this way, Timescale breaks bigger tables in an unspecified number of smaller tables, doing the same for also for indexes. So each operation can be broken up in smaller operations, allowing the parallelization, and resulting in better performances. Of course it is harder to handle such structure, since each operation made on a table could involve an Hypertable. In this case, to make each chunk to behave like a normal table, it is necessary to split the operation transparently, resulting in a harder query planning (Fig. 3).

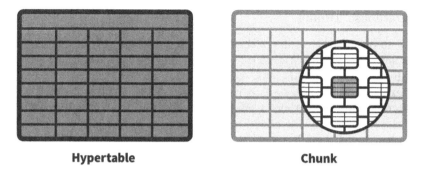

Hypertable   **Chunk**

Fig. 3. Hypertable and Chunk from official documentation [21]

3.1 The Investigated Settings

Our first idea was to understand the benefits and draw-backs of the introduction of a TSMS like Timescale, in terms of performances, making rise to two settings: *PostgreSQL + PostGIS* vs. *PostgreSQL + PostGIS + Timescale.*

On the other hand, from the above presentation of these solutions, we can see that one of the main differences inducted by the use of Timescale is a different ordering of the records within the physical files on the mass storage. This motivated us to investigate whether differences in performances (if any) might be due to the new algorithms and data structures of the TSMS, or simply to the different physical ordering of the records. As a consequence, to better understand the differences among the mentioned technologies, a further setting has been defined, exploiting the possibility offered by *PostgreSQL* to specify a custom order of the records in the files. Thus, in this setting, the table *parking_events*, containing massive temporal data, has been **clustered**, i.e. physically ordered, on the basis of the *ArrivalTime* attribute.

As a result, we compared the performances achievable by the three following off-the-shelf data management settings:

1. The Relational DBMS *PostgreSQL* with the spatial extension *PostGIS*, using the indexing solutions offered by the DBMS. We will refer to this as *Setting 1: PostgreSQL and PostGIS.*
2. The Relational DBMS *PostgreSQL* with the spatial extension *PostGIS*, using advanced data sorting and indexing capabilities of the DBMS. We will refer to this as *Setting 2: Clustered PostgreSQL and PostGIS*
3. The Relational DBMS *PostgreSQL* with the spatial extension *PostGIS* and with the time-series management extension Timescale, a kind of storage solution specialized in handling temporal data. We will refer to this as *Setting 3: Timescale and PostGIS*

3.2 Database Tuning

In order to present the results of the investigated settings, it is first necessary to understand some peculiarities of the data indexing we used for our experiments, especially for the temporal part.

Indexing methods are used to increase the speed of data retrieval. In the spatio-temporal context, there are non-trivial issues in indexing, due to the fact that data continuously increases over the time [37]. The choice of the right indexes is a key part of database tuning. A lack of suitable indexes will lead to the scan of the entire files from the disk during the execution of a query, with disastrous impacts on the performances. On the other hand, too much indexes will slow down data ingestion, as well as will occupy significant space on the disk.

In our case, we defined indexes as follows. In all the three settings, the table *bay_locations*, not presenting temporal dimensions, has two indexes. The first is a *B*-tree on the *meter_id* attribute, used to search locations by the ID of the sensor. The second one is an R-Tree on the attribute *the_geom*, used to speed-up the spatial queries. The *parking_events* table, containing the massive amount of temporal data, is handled differently in each of the three settings. In *Setting 1*, a *B*-tree index has been defined on the *ArrivalTime* attribute, to support time-based queries. Since this attribute is not primary key, the records in the files on the storage memory are not physically ordered by this field. As a consequence, the *B*-tree index on *ArrivalTime* is a secondary, dense index. Another *B*-tree has been defined on the *StreetMarker* attribute, which is a Foreign Key for the *bay_locations* table, to optimize the join. In *Setting 2*, the attribute *ArrivalTime* becomes the physical ordering field of the records, and thus the corresponding index becomes clustered. Finally, in *Setting 3*, we converted the *parking_events* table in an *Hypertable*, optimized for Time Series filtering, with a default interval dimension of 1 week. Being mandatory for each Hypertable, chunks are both internally and externally physically ordered by the time index (i.e. *ArrivalTime*), improving both range and specific value searches.

In Table 1 we summarize the characteristics of the three investigated settings.

Table 1. Characteristics of the investigated settings

Setting	Technologies	Characteristics
1	PostgreSQL, PostGIS	B-Tree, R-Tree (GiST)
2	PostgreSQL, PostGIS	B-Tree, R-Tree (GiST), B-Tree clustered by time
3	PostgreSQL, PostGIS, Timescale	B-Tree, R-Tree (GiST), natively ordered by time, hypertable

4   The Empirical Evaluation

In this section we describe and discuss the empirical evaluation of three different settings, including details on the evaluation protocol and the results.

4.1   Experimental Setup

In order to assess the consequences of the introduction of a TSMS on top of a GIS for managing massive mobility data, we empirically compared the performances achievable by the three settings on some queries, using the complete Melbourne dataset, i.e. 275 million records. As done in some works on DBMS benchmarking (e.g.: [15]), our experimental protocol aims at the assessment of a specific performance indicators, evaluated in some typical situations. In this case the most used indicator is the *Retrieval Time* (see Table 2) in seconds. We investigated many spatio-temporal queries, obtaining homogeneous results. Thus, for sake of brevity, we will describe in details only one query, reported in listing 1.1, meant to represent typical data retrievals that a Decision Maker of a Smart City might perform, to get a better vision on mobility phenomena. Such location-based query is designed to retrieve all the parking events that happened in a specific *Range* from a given *Center*, in a time interval going from *StartInterval* to *EndInterval*.

Listing 1.1. The parametric Spatio-temporal query we used in our experiments.

```
SELECT * FROM parking_events AS PE JOIN bay_locations AS BL
    ON PE.StreetMarker = BL.marker_id
WHERE PE.ArrivalTime >= StartInterval
    AND PE.ArrivalTime < EndInterval
    AND PE.DepartureTime < EndInterval
    AND st_dwithin(BL.the_geom, Center, Range)
```

To evaluate the proposed settings, we fixed the *center* of each spatio-temporal query in *Lonsdale Street*, then we varied spatial and temporal parameters. In particular, the range parameter varies among 100, 500 and 1000 m, while the length of time period varies among 1,2,3,7,14 and 21 days.

In the results we will report the average of five runs of the spatio-temporal query reported in listing 1.1, for each combination of the mentioned parameters and each of the three settings. The machine on which the experiments were executed is a Dell workstation, equipped with a fourth generation quad-core Intel i7 processor, 24 GB RAM and Crucial SSD 500 GB, SATA3.

Finally, another key metrics for the effectiveness of a Data Management solution is the *Disk Occupancy* (see Table 2) of the dataset, including also the required indexes.

Table 2. Selected evaluation metrics

Metrics	
Retrieval Time	Time required to retrieve and return records (in seconds)
Disk Occupancy	Space required to store data and indexes (in Gygabytes)

4.2 Results and Discussion

In Figs. 4, 5, and 6, we report plots of the results achieved by each setting and each parameter combination, in seconds.

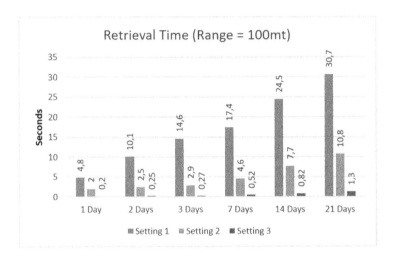

Fig. 4. Average execution time for queries with range of 100 m, in Seconds

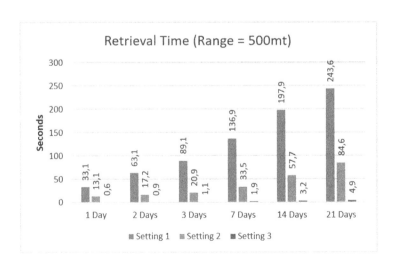

Fig. 5. Average execution time for queries with range of 500 m, in Seconds

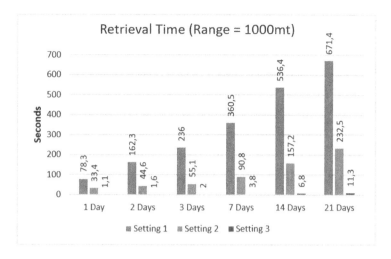

Fig. 6. Average execution time for queries with range of 1000 m, in Seconds

As we can see from these Figures, the improvement in terms of performances inducted by the introduction of Timescale is impressive. The results in terms of required seconds to run the query in Setting 3 are on average 97% lower than Setting 1. The average execution of the query on a range of 100 m and 21 days took 30.7 s with *PostgreSQL* and *PostGIS*, and only 1.3 s on *Timescale*. Moving to a 1000 m range, the difference in time is 671.4 s vs. just 11.3.

Also Setting 2 represent a significant improvement over Setting 1, so the physical ordering of the records has a deep impact on the execution time, as expected. On average, Setting 2 reduces the time required to run the query of about 62%. Nevertheless, the performances of Setting 3 are by far better than Setting 2, so the optimization of TimeScale can make an important difference.

In Fig. 7 we report the requirements of the three Settings in terms of disk occupancy. As we can see, the Setting 3 needs a total of 54 GB, that is 2 GB more than Settings 1 and 2. This is due to the table partitioning in chunks of one week, operated by Timescale.

Looking at these results, it is clear that the physical ordering is a key factor to obtaining good performances in retrieval tasks over massive data. It is also important to note that, when we imported the datasets to populate the DBMS for Setting 2, the sorting of the records required about 79 min and must be performed after every update on the ordering attribute. On the other hand, Timescale act like an instance of PostgreSQL in continuous sorting of the records, splitting the overhead of the ordering phase in the ingestion phase on smaller time-ordered tables. This way Timescale guarantee a collection of physically time-ordered tables, rapidly searchable and able to completely suit in memory. As a minor drawback, Timescale requires that some parameters must be set (i.e. the *chunk_time_interval*), in order to achieve and maintain such performances, without an excessive disk usage increment.

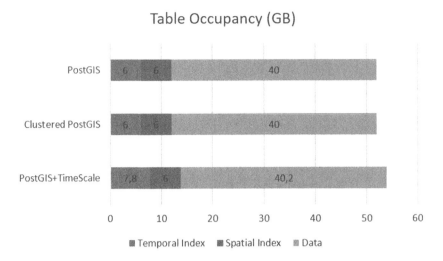

Fig. 7. Disk occupancy of time data table.

5   Conclusions

Smart Cities are an evolving concept, which needs for a series of enabling technologies in order to take advantage of all the actual data sources like IoT sensors, smartphones, probe vehicles and so on. To do so, we must be able to handle and process, rapidly and in an efficient way, huge quantities of heterogeneous data, with high spatial and temporal resolution.

In this study we reported an empirical experience in handling a massive spatio-temporal mobility dataset, covering about seven years of on-street parking availability, from the municipality of Melbourne (AU). In particular, we compared a consolidated setting (i.e. PostgreSQL and PostGIS) to manage georeferred data, with an optimized version of it (i.e. clustered) and with a third method based on an advanced Time Series Management System (i.e. Timescale), providing some techniques optimized to manage massive temporal data in the modern technological context.

The empirical results on a complex spatio-temporal query, involving multiple location-based and temporal parameters, show that Timescale is able to outperform by far any other option. On the other hand, it requires more space on the disk and some additional parameters to be set up at the creation of the Database. Moreover, the possibility to specify in PostgreSQL the sorting attribute may also impact the results in a dramatic way.

As future evolution, we are interesting in evaluating the effectiveness of the proposed solution in an on-line fashion with different data sources and also with a stream of incoming data. Another interesting aspect would be the comparison with DBMS based on non-relational data models, like the graph-based ones, which also support some kind of spatial primitives.

References

1. Jin, J., Gubbi, J., Marusic, S., Palaniswami, M.: An information framework for creating a smart city through Internet of Things. IEEE Internet Things J. **1**(2), 112–121 (2014)
2. Sun, Y., Song, H., Jara, A.J., Bie, R.: Internet of Things and big data analytics for smart and connected communities. IEEE Access **4**, 766–773 (2016)
3. Bélissent, J., et al.: Getting clever about smart cities: new opportunities require new business models, Cambridge, Massachusetts, USA, vol. 193, pp. 244–277 (2010)
4. Atzori, L., Iera, A., Morabito, G.: The Internet of Things: a survey. Comput. Netw. **54**(15), 2787–2805 (2010)
5. Bock, F., Di Martino, S., Origlia, A.: Smart parking: using a crowd of taxis to sense on-street parking space availability. IEEE Trans. Intell. Transp. Syst. **21**(2), 496–508 (2020)
6. Kwoczek, S., Di Martino, S., Nejdl, W.: Stuck around the stadium? An approach to identify road segments affected by planned special events. In: 2015 IEEE 18th International Conference on Intelligent Transportation Systems, pp. 1255–1260. IEEE (2015)
7. Zhang, J., Wang, F.Y., Wang, K., Lin, W.H., Xu, X., Chen, C.: Data-driven intelligent transportation systems: a survey. IEEE Trans. Intell. Transp. Syst. **12**(4), 1624–1639 (2011)
8. Bock, F., Martino, S.D., Sester, M.: What are the potentialities of crowdsourcing for dynamic maps of on-street parking spaces? In: Proceedings of the 9th ACM SIGSPATIAL International Workshop on Computational Transportation Science, pp. 19–24 (2016)
9. Chen, M., Mao, S., Zhang, Y., Leung, V.C.M.: Big Data: Related Technologies, Challenges and Future Prospects. SCS. Springer, Cham (2014). https://doi.org/10.1007/978-3-319-06245-7
10. Kitchin, R.: The real-time city? Big data and smart urbanism. GeoJournal **79**(1), 1–14 (2014)
11. Chen, W., Guo, F., Wang, F.Y.: A survey of traffic data visualization. IEEE Trans. Intell. Transp. Syst. **16**(6), 2970–2984 (2015)
12. Ferreira, N., Poco, J., Vo, H.T., Freire, J., Silva, C.T.: Visual exploration of big spatio-temporal urban data: a study of New York city taxi trips. IEEE Trans. Visual. Comput. Graphics **19**(12), 2149–2158 (2013)
13. Compieta, P., Di Martino, S., Bertolotto, M., Ferrucci, F., Kechadi, T.: Exploratory spatio-temporal data mining and visualization. J. Vis. Lang. Comput. **18**(3), 255–279 (2007)
14. Jensen, S.K., Pedersen, T.B., Thomsen, C.: Time series management systems: a survey. IEEE Trans. Knowl. Data Eng. **29**(11), 2581–2600 (2017)
15. Di Martino, S., Fiadone, L., Peron, A., Riccabone, A., Vitale, V.N.: Industrial Internet of Things: persistence for time series with NoSQL databases. In: 2019 IEEE 28th International Conference on Enabling Technologies: Infrastructure for Collaborative Enterprises (WETICE), pp. 340–345. IEEE (2019)
16. Robino, C., Di Rocco, L., Di Martino, S., Guerrini, G., Bertolotto, M.: A visual analytics GUI for multigranular spatio-temporal exploration and comparison of open mobility data. In: 22nd International Conference Information Visualisation (IV), pp. 309–314. IEEE (2018)
17. Kwoczek, S., Di Martino, S., Rustemeyer, T., Nejdl, W.: An architecture to process massive vehicular traffic data. In: 2015 10th International Conference on P2P, Parallel, Grid, Cloud and Internet Computing (3PGCIC), pp. 515–520. IEEE (2015)

18. Makris, A., Tserpes, K., Anagnostopoulos, D., Nikolaidou, M., de Macedo, J.A.F.: Database system comparison based on spatiotemporal functionality. In: Proceedings of the 23rd International Database Applications and Engineering Symposium, p. 21. ACM (2019)

19. PostgreSQL: PostgreSQL The Most Advanced Open-source Object Relational Database

20. PostGIS: PostGIS spatial database extender for PostgreSQL (2019). https://postgis.net/docs/. Accessed 15 December 2019

21. Timescale: Timescale simple, scalable SQL for time-series and IoT (2019). https://www.timescale.com/. Accessed 15 December 2019

22. SFMTA: SFPark: Putting Theory Into Practice. Pilot project summary and lessons learned (2014). Accessed 24 June 2016

23. Bock, F., Di Martino, S.: On-street parking availability data in San Francisco, from stationary sensors and high-mileage probe vehicles. Data Brief **25**, 104039 (2019)

24. Government, V.S.: City of Melbourne. Melbourne parking sensor (2014). https://www.melbourne.vic.gov.au/SiteCollectionDocuments/parking-technology-map.pdf. Accessed 15 December 2019

25. Zheng, Y.: Trajectory data mining: an overview. ACM Trans. Intell. Syst. Technol. (TIST) **6**(3), 29 (2015)

26. Naqvi, S.N.Z., Yfantidou, S., Zimányi, E.:Time series databases and influxdb. Studienarbeit, Université Libre de Bruxelles (2017)

27. Kaur, K., Rani, R.: Managing data in healthcare information systems: many models, one solution. Computer **48**(3), 52–59 (2015)

28. Liu, X., Nielsen, P.S.: A hybrid ICT-solution for smart meter data analytics. Energy **115**, 1710–1722 (2016)

29. Gilbert, S., Lynch, N.: Perspectives on the CAP theorem. Computer **45**(2), 30–36 (2012)

30. Zhang, L., Yi, J.: Management methods of spatial data based on PostGIS. In: 2010 Second Pacific-Asia Conference on Circuits, Communications and System, vol. 1, pp. 410–413. IEEE (2010)

31. Consortium, O.G., et al.: OpenGIS implementation specification for geographic information-simple feature access-part 2: SQL option. OpenGIS Implementation Standard (2010)

32. Guttman, A.: R-trees: a dynamic index structure for spatial searching, vol. 14. ACM (1984)

33. Papadias, D., Tao, Y., Kanis, P., Zhang, J.: Indexing spatio-temporal data warehouses. In: Proceedings 18th International Conference on Data Engineering, pp. 166–175. IEEE (2002)

34. Lehman, P.L., et al.: Efficient locking for concurrent operations on B-trees. ACM Trans. Database Syst. (TODS) **6**(4), 650–670 (1981)

35. Hellerstein, J.M., Naughton, J.F., Pfeffer, A.: Generalized search trees for database systems, September 1995

36. Michel, O., Sonchack, J., Keller, E., Smith, J.M.: PIQ: Persistent interactive queries for network security analytics. In: Proceedings of the ACM International Workshop on Security in Software Defined Networks and Network Function Virtualization, pp. 17–22. ACM (2019)

37. John, A., Sugumaran, M., Rajesh, R.: Indexing and query processing techniques in spatio-temporal data. ICTACT J. Soft Comput. **6**(3), 1198 (2016)

The Integration of OGC SensorThings API and OGC CityGML via Semantic Web Technology

Yao-Hsin Chiang[1,3], Chih-Yuan Huang[2(✉)], and Masaaki Fuse[3]

[1] Department of Civil Engineering, National Central University, Taoyuan 320, Taiwan
[2] Center for Space and Remote Sensing Research, National Central University, Taoyuan 320, Taiwan
cyhuang@csrsr.ncu.edu.tw
[3] Graduate School of Engineering, Hiroshima University, 1-4-1 Kagamiyama, Higashi-Hiroshima, Hiroshima 739-8527, Japan

Abstract. Smart cities effectively integrate human, physical, and digital systems operating in the built environment to provide automatic and efficient applications. While 3D city models, Internet of Things (IoT), and domain models are essential components of smart cities, the integration of IoT resources and 3D city models is a central information backbone for smart city cyber-infrastructures. However, we argue that most of the existing solutions integrating 3D city models and IoT resources are usually customized according to applications and lack of interoperability. To improve the interoperability between smart city modules, this study proposes a semantic-based methodology to integrate OGC CityGML and SensorThings API standards. Based on the data models from these two standards, this study proposes an integration ontology to connect information from these two standards. Due to the flexible definition of Thing in the IoT, the proposed ontology also considers multiple views of Thing. As a result, information from the CityGML and SensorThings API can be connected and queried via SPARQL queries. In general, this proposed integration ontology can facilitate the integration of IoT and 3D city models to achieve interoperable smart city.

Keywords: Open standard · Semantic web · CityGML · SensorThings API

1 Introduction

In recent years, more than half of the world's population lives in cities, and the increase in population has put tremendous pressure on the environment and the structure of the city. With the advancement of technology, the concept of smart cities has attracted attentions from various fields [1, 2], where many smart city services require spatial information [3]. Spatial information management and processing technologies are necessary components in a smart city infrastructure, such as the Global Navigation Satellite System, Building Information Modelling, Location-Based Service, etc. On the other hand, to capture the dynamics of a city and perform real-time actions, sensing and tasking

© Springer Nature Switzerland AG 2020
S. Di Martino et al. (Eds.): W2GIS 2020, LNCS 12473, pp. 55–67, 2020.
https://doi.org/10.1007/978-3-030-60952-8_6

capabilities supported by the IoT (Internet of Things) [4] is also an essential smart city component. However, spatial information and IoT resources are formulated as individual datasets/services following different data model or protocol standards, where the linkage between these two types of resources has not received enough attention. Therefore, this study focuses on the integration of spatial information and the IoT resources.

In fact, in recent years, many studies have begun to discuss the integration between 3D model spatial information and the IoT [5]. To achieve effective integration, interoperability between these two types of data must be considered. As open standards are the foundation to achieve interoperability [6], the integration of 3D city models and IoT services should also follow standards. Among the existing standards, this research selects the OGC (Open Geospatial Consortium) CityGML [7] and SensorThings API [8] as they define comprehensive and semantic-rich data models for 3D model spatial information and IoT resources, respectively.

Regarding the integration method, this research utilizes the semantic web technology [9] to connect the semantic of 3D model and IoT resources [10]. The semantic web technology can describe relationships between two entities represented as URIs. As relationships can be easily extended, the semantic web is flexible to describe various relationships and can maintain the independence of data sources. In order to integrate the 3D model and IoT resources, this study proposes an integration ontology that defines the relationships between the CityGML and SensorThings API data models. In addition, as the definition of "Thing" in the IoT is flexible, ontologies for different views of the Thing are formulated to support various use cases.

Based on the proposed ontology, the linkage between these 3D model and IoT resources are described in the RDF (Resource Description Framework) format [11] that can be easily queried through SPARQL queries to support various applications. For instance, from the 3D model perspective, the proposed solution can help retrieve dynamic sensing information of a space, a door, a window, such as temperature observations, relative humidity readings, open/close status, locked/unlocked status, etc. Applications can also invoke the IoT tasking capabilities to remotely control physical instruments, such as street lights, door locks, air conditions, etc. On the other hand, from the IoT perspective, sensor observations and tasking capabilities from different IoT devices can be analyzed and queried according to different spatial relationships, such as within the same room, connecting two spaces, etc.

In general, the proposed ontology can effectively integrate 3D model and IoT resources to support the realization of smart city applications. In the next section, we review related literatures and 3D model and IoT standards. Section 3 introduces the proposed integration ontology to connect the CityGML and SensorThings API. In Sect. 4, we present some preliminary results. Finally, Sect. 5 concludes this research.

2   Related Works

2.1   Integration of 3D Model and IoT Resources for Smart City

In pursuit of the smart city vision, smart cities should integrate cross-disciplinary components [12]. Many studies have discussed the integration of IoT sensors with 3D models. For example, Wang et al. [13] introduce the integration of indoor sensor information and indoor route network information to enable dynamic risk assessment of evacuation for immediate evacuation route planning. Wang et al. [14] attempted to integrate the BIM (Building Information Model) house model and the OGC SOS (Sensor Observation Service) web service at the application layer in the three-layer network architecture (i.e., data repository, data service, applications) to retrieve observations in each room. Chaturvedi et al. [15] achieved the solar radiation assessment of building surfaces and roofs by integrating CityGML and SOS services, as well as historical observations of solar energy. However, many existing researches either did not consider 3D model and IoT standards or merely integrate resources according to different customized applications, where a general integration solution was not discussed.

2.2   Open Standards for the IoT Resources

Due to the severe heterogeneity issue of IoT resources, data models and communication protocols for the IoT should be unified with open standards [16]. Open Connectivity Foundation (OCF) defines the connectivity requirements to improve interoperability between devices [17]. Industrial Internet Consortium (IIC) proposes Industrial Internet Connectivity Framework (IICF), which is used to provide a reference framework for Industrial Internet of Things (IIoT) system connection [18]. The oneM2M [19] defines a horizontal M2M service platform which aims to unify communication between devices and services. From the Sensor Web field, the OGC Sensor Observation Service (SOS) [20] defines a web service protocol standard for sensor resource sharing.

The OGC SensorThings API [8, 21] data model is mainly based on the OGC Observation and Measurement (O&M) [22] and Sensor Model Language (SensorML) [23] standards and its web service protocol is inspired by the Organization for the Advancement of Structured Information Standards (OASIS) Open Data Protocol (OData) standard [24]. Comparing to other open IoT service standards, the OGC SensorThings API defines a complete and versatile data model, and its web service interface provides highly flexible data query capabilities with the Representational State Transfer (RESTful) style and JavaScript Object Notation (JSON) encoding [8]. To support complete IoT functionalities, the OGC SensorThings API part 1 and part 2 standards [8, 21] define the data models and service protocols for the sensing and tasking capabilities, respectively. In general, we believe that the OGC SensorThings API is one of the most complete IoT web service standards.

2.3   Open Standard for the 3D Model

Among the 3D model standards, the OGC CityGML standard [7] is a widely accepted data model for the representation and exchange of 3D city and landscape models comprising information on geometry, topology, appearance, and semantics of the most common urban feature. The CityGML has several advantages. For instance, multi-scale modeling is one of the core features of CityGML, defining five Level of Detail (LoD) to represent city models for different application scenarios. This standard also defined a complete and semantic-rich data model to describe model information in a precise and interoperable manner.

Some studies try to integrate CityGML with other standards. For example, Vilgertshofer et al. [25] combined the data models from the CityGML LoD 4 and BIM based on a linked data approach. Kim et al. [26] proposed a methodology to formulate IndoorGML from CityGML to achieve indoor navigation applications. Chaturvedi et al. [15] designed an attribute in the CityGML called Dynamizer, which attaches sensor observations or database connections to the static 3D model for dynamic properties. In general, we believe that the CityGML is one of the most comprehensive solutions that support various smart city applications from city-scale to indoor-scale.

3   Methodology

This research focuses on integrating the OGC CityGML and SensorThings API to link geospatial information and IoT capabilities to facilitate smart city applications. In this section, we introduce the proposed integration method.

3.1   OGC CityGML

The OGC CityGML is an international open standard for the representation of 3D city model information. The CityGML can represent the geometrical, semantic, and visual aspects of 3D city models. The CityGML data model is semantic-rich and complete enough to describe various features in the city (e.g., wall, roof, door, window, room, etc.) and also some thematic modules (e.g., building, tunnel, bridge, etc.). Furthermore, the CityGML defines five LoD, from LoD 0 to LoD 4 for different applications. In this research, we discuss the relationship between the classes of CityGML LoD 4 and SensorThings API.

3.2   OGC SensorThings API

The OGC SensorThings API is an IoT web service open standard. It defines several classes and attributes as the data model for IoT resources, including the sensing and tasking capabilities. Currently, there are two parts of SensorThings API standard, where part 1 for sensing capability and part 2 for tasking capability. A brief explanation of each data model class is introduced as follows.

- Thing: A Thing is an object of the physical world (physical Things) or the information world (virtual Things) that is capable of being identified and integrated into communication networks [27].
- Location: The entity recording Thing's last known location.
- HistoricalLocation: Record the time points or time periods of the previous locations of a Thing. For instance, if the thing is mobile, there should be several Locations link to HistoricalLocations.
- Datastream: A Datastream groups a collection of Observations measuring the same ObservedProperty and produced by the same Sensor. For example, if a Thing can observe three phenomena, e.g., air temperature, relative humidity, and illumination, this Thing may have three Datastreams, where each one groups Observations of one phenomenon.

- `Sensor`: An instrument that observes a property or phenomenon.
- `ObservedProperty`: The phenomenon of an observation.
- `Observation`: The act of measuring or determining the value of a property. An `Observation` in SensorThings API represents a single `Sensor` reading of `ObservedProperty`.
- `FeatureOfInterest`: The observed feature of an `Observation`.
- `TaskingCapability`: The controllable capabilities that is supported by a `Thing`.
- `Task`: This class is a user command for controlling a `TaskingCapability`. The device should be controlled according to input values specified in the `Task`.
- `Actuator`: This class introduces the metadata of the instrument supporting the `TaskingCapability`.

SensorThings API applies JSON format and RESTful web service style to host IoT resources. A `Thing` contains several attributes such as name, description, properties. In addition to these attributes, SensorThings API also contains navigation links connecting to related entities.

In general, the SensorThings API provides a complete solution for the IoT web service layer. This standard defines a comprehensive and general data model for IoT sensing and tasking capabilities. Furthermore, SensorThings API provides RESTful style web service protocol and flexible query functions for users to query targeted IoT resources.

3.3 Integration Strategy

As mentioned earlier, this research applies the semantic web technology to integrate the CityGML and SensorThings API data models. The semantic web technology is an effective and flexible solution to describe various relationships between data from different domains.

One of the unique challenges in the integration between 3D model and IoT resources is that the `Thing` in the IoT could have different definitions depending on different applications. For instance, a `Thing` could be a room, a door, a window, a corridor, a sensor, an appliance, etc. In this case, the connection between 3D model and IoT resources should consider scenarios that follow different `Thing` definitions.

As the semantic web can connect resources by linking URIs (Uniform Resource Identifier) together and describe their relationships, data can be generated and represented independently in different storages. To apply the semantic web technology to integrate 3D model and IoT resources, an integration ontology that defines the relationships between CityGML and SensorThings API data models is the most important component. To be specific, this research defines the integration ontology with OWL (Ontology Web Language). Based on the ontology, we can describe the relationships between 3D model and IoT resources in the RDF format, which can be queried by SPARQL queries.

3.3.1 Resource Properties for Integration Ontology Framework

In the integration ontology proposed by this research, sosa[1] (Semantic Sensor Network Ontology) published by the W3C (World Wide Web Consortium) in 2017 is applied to describe the relationships for `Datastream` and `TaskingCapability` (i.e., *hosts* and *observes*). In addition, the CityGML ontology[2] has already done by [32]. Regarding the SensorThings API ontology, we construct the STA-Lite that includes the necessary relationships between SensorThings API classes, i.e., *hasDatastream*, *hasTaskingCapability*, and *hasObservedProperty*. For relationships that cannot be found in existing ontologies, we define those relationships by ourselves, and represent with a namespace *cgis*, which are *isThing*, *happensIn*, *isProvidedBy* and *isProducedBy*.

3.3.2 Multiple Definitions of the IoT Thing

This research proposes an integration ontology to describe the relationship between the classes in the CityGML and SensorThings API data models. Considering the multiple definitions of the IoT `Thing`, we design the ontology framework from different views, including a-building-as-a-Thing (Fig. 1), a-room-as-a-Thing (Fig. 2), an-opening-as-a-Thing (Fig. 3) and a-device-as-a-Thing (Fig. 4).

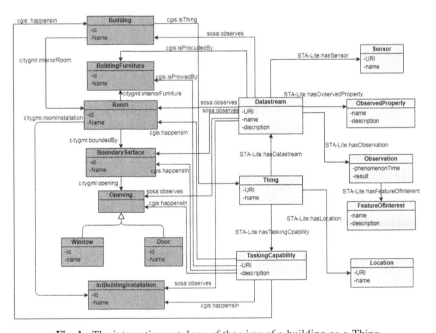

Fig. 1. The integration ontology of the view of a-building-as-a-Thing

---

[1] https://www.w3.org/TR/vocab-ssn/#SOSA.
[2] http://cui.unige.ch/isi/onto//citygml2.0.owl.

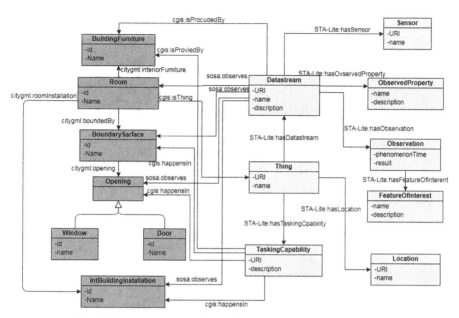

Fig. 2. The integration ontology of the view of a-room-as-a-Thing

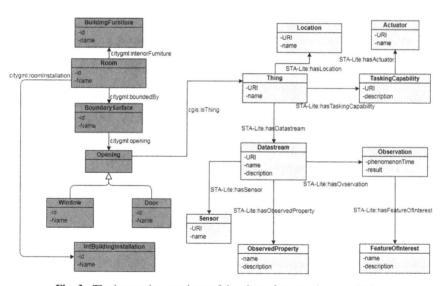

Fig. 3. The integration ontology of the view of an-opening-as-a-Thing

To help the explanation in this paper, we use the `Courier New` font style for SensorThings API classes and entities, < ... > to represent CityGML resources and the *Italic* font style for the relationship (i.e., predicate).

Figure 1 shows the view of a-building-as-a-Thing. In this situation, a building contains several components (e.g., <Room>, <BuildingFurniture>, <Opening>, etc.).

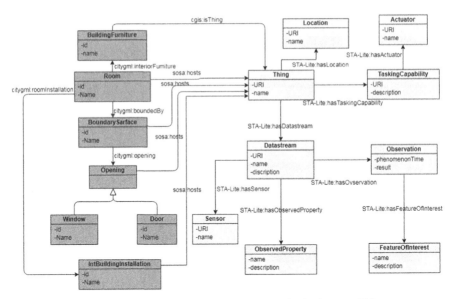

Fig. 4. The integration ontology of the view of a-device-as-a-Thing

The CityGML classes are shown in blue and SensorThings API classes are shown in yellow. As the concept that a building contains several rooms and components is similar with a-room-as-a-Thing view, where a `Thing` contains several entities. In this view, a `Datastream` of the `Thing` *observes* a phenomenon that happens in the building, which may observe the room or any component in the building. On the other hand, a `TaskingCapability` *happensIn* a specific entity in the building. The relationships of `Datastreams` and `TaskingCapability` in this view are similar with the view of a-room-as-a-Thing. The `Datastream` can be produced by <BuildingFurniture> and the `TaskingCapability` could also be provided by <BuildingFurniture> .

The view of a-room-as-a-Thing is shown in Fig. 2. In this view, a `Datastream` of the `Thing` *observes*[3] a phenomenon that happens in the room, which may observe the room or any component in the room. A `Datastream` is the collection of `Observations` according to an `ObservedProperty` of the room. On the other hand, a `TaskingCapability` *happensIn* a specific entity. In this view, `Thing` might contain various elements (e.g., thermometer, buzzer, etc.), which represent the feature of the phenomenon or status of the room. Those features are capable of sensing and tasking capability happens in the room. A <BuildingFurniture> entity *provides* `TaskingCapabilities` (e.g., change of illumination, turn on/off alarm, etc.) and *produces* `Datastreams` (e.g., air temperature, relative humidity, etc.).

The view of an-opening-as-a-Thing is shown by Fig. 3. An <Opening> element including the <Window> and the <Door> is represented as a `Thing`. In this view, the `Datastream` of the `Thing` can record the status of the <Opening> (e.g. opening/closing) or the condition (e.g. the size of passage). The `TaskingCapability`

---

[3] https://www.w3.org/TR/vocab-ssn/#SOSAobserves.

describing the information of the actuator embedded in the door or window that supports controllable actions. In addition, the relationships between the `Thing` and other CityGML entities can also be described based on the proposed ontology.

Figure 4 shows the view of a-device-as-a-Thing. In this situation, we map the `Thing` in SensorThings API to the <BuildingForniture> of CityGML. If the CityGML has a resource directly corresponding to the `Thing`, the CityGML resource and the `Thing` can be linked directly. However, if there is no corresponding <BuildingForniture> resource in CityGML and only a containing feature is presented, the integration ontology also supports the relationships between the `Thing` and the CityGML containing feature, such as <Room>, <BoundarySarface>, <Opening>, and <IntBuildingInstallation>, which *hosts*[4] the `Thing`.

The integration ontology is recorded in the RDF format. Based on the ontology, 3D model and IoT resources can be linked and represented in the RDF format as well. In addition, due to the flexibility of the semantic web technology, relationships from multiple views of Thing can be applied in the same dataset, which help support more smart city applications.

4 Experimental Result

To demonstrate the preliminary result of the proposed solution, we implemented some use cases that require queries linking 3D model and IoT resources. Based on the integration ontology, we can retrieve data via SPARQL queries.

4.1 Testing Dataset

For the 3D city model, we use an open-source dataset[5] of CityGML LoD 4 (Fig. 5). There are 277 features in the model with complete information based on the CityGML data model. We simulated some `Thing` entities, which correspond to features in the 3D model. Although the `Things` were simulated, we have created corresponding `Things` in a SensorThings API service supporting sensing and tasking capabilities. Furthermore, based on the proposed ontology, the connections between these two datasets were described in the RDF format.

4.2 Query Integrated Data

In order to demonstrate the integration of the CityGML dataset and SensorThings API service, Fig. 6 shows the SPARQL query that requesting the element information of <Room> elements from CityGML and their corresponding `Thing`'s URIs from SensorThings API. The result is shown in Fig. 7. With `Thing` URIs, we can retrieve the `Datastreams` from the SensorThings API web service. For example, by querying the URI (*STA/v1.0/Thing(16)/Datastreams*). Furthermore, as a SPARQL query (Fig. 8) result shown in Table 1, a `TaskingCapability` is related to a CityGML <IntBuildingInstallation> which is installed at Room_002.

---

[4] https://www.w3.org/TR/vocab-ssn/#SOSAhosts.

[5] http://www.citygmlwiki.org/index.php?title=KIT_CityGML_Examples.

Fig. 5. The CityGML LoD4 dataset used in this study

```
PREFIX cigs: <http://140.115.110.71/cgis.owl#>
PREFIX STA-Lite: <http://140.115.110.71/STA-Lite.owl#>
PREFIX citygml2: <http://cui.unige.ch/isi/onto//citygml2.0.owl#>
PREFIX rdf: <http://www.w3.org/1999/02/22-rdf-syntax-ns#>
PREFIX FJK_link: <http://140.115.110.71/FJK_link.owl#>

select DISTINCT ?Room_Name ?Room_ID ?Thing_URI
where {
        ?room rdf:type citygml2:Room .
        ?room FJK_link:id ?Room_ID .
        ?room FJK_link:Name ?Room_Name .
        ?room cgis:isThing ?Thing .
        ?Thing STA-Lite:URI ?Thing_URI
}
```

Fig. 6. An example of using SPARQL query for the integration ontology

Room_Name ⬦	Room_ID ⬦	Thing_URI
1 001	GMLID_BUI6147_355_7717	'http://140.115.110.69:8080/_sta_taskingservice_/STA/v1.0/Things[6]'
2 002	GMLID_BUI253135_1424_3471	'http://140.115.110.69:8080/_sta_taskingservice_/STA/v1.0/Things[7]'
3 003	GMLID_BUI324751_1122_1743	'http://140.115.110.69:8080/_sta_taskingservice_/STA/v1.0/Things[8]'
4 004	GMLID_BUI98157_1592_9125	'http://140.115.110.69:8080/_sta_taskingservice_/STA/v1.0/Things[9]'
5 005	GMLID_BUI373595_277_13251	'http://140.115.110.69:8080/_sta_taskingservice_/STA/v1.0/Things[10]'
6 006	GMLID_BUI29461_942_4450	'http://140.115.110.69:8080/_sta_taskingservice_/STA/v1.0/Things[1]'
7 101	GMLID_BUI38955_1837_8592	'http://140.115.110.69:8080/_sta_taskingservice_/STA/v1.0/Things[2]'

Fig. 7. A SPARQL result of the corresponding elements of CityGML and SesnsorThings API

4.3   Demonstration of Supporting Multiple Views of Thing

The proposed integration ontology can describe different definitions of `Thing`. For example, an IoT door lock embedded in a door is represented as that a CityGML <Door> element contains a `TaskingCapability` based on the an-opening-as-a-Thing view. In the meantime, the door lock can also be recognized as a device hosted by the <Door> element based on the a-device-as-a-Thing view. Figure 9 and Fig. 10 demonstrate the

```
select DISTINCT ?TaskingCapability_URI ?CityGMLobj_Name ?type_of_x ?room
where {
    ?Thing STA-Lite:hasTaskingCapability ?TaskingCapability .
    ?TaskingCapability STA-Lite:URI ?TaskingCapability_URI ;
                        cgis:happensIn ?x .
    ?x FJK_link:Name ?CityGMLobj_Name ;
        rdf:type ?type_Of_x .
    ?room citygml2:roomInstallation ?x
}
```

Fig. 8. An example of using SPARQL query for the integration ontology

Table 1. A SPARQL result of query shown in Fig. 8

TaskingCapability_URI	CityGMLobj_Name	Type_Of_x	Room
http://140.115.110.69:8080/_sta_taskingservice_/STA/v1.0/TaskingCapabilities(1)	Heizung	citygml2:IntbuildingInstallation	Room_002

SPARQL queries to request for the door lock information from different views of `Thing`. As the query results are the same (Table 2), it demonstrates that different views of `Thing` can be supported in the integration ontology.

```
select DISTINCT ?door_name ?id ?Tasking_URI
where {
    ?door rdf:type citygml2:Door ;
        FJK_link:Name ?door_name ;
        FJK_link:id ?id ;
        cgis:isThing / STA-Lite:hasTaskingCapability ?Tasking .
    ?Tasking STA-Lite:URI ?Tasking_URI .
}
```

Fig. 9. A SPARQL query example of a door lock in the an-opening-as-a-Thing view

```
select DISTINCT ?Name ?id ?Tasking_URI
where {
    ?x sosa:hosts ?Thing ;
        FJK_link:Name ?Name ;
        FJK_link:id ?id .
    ?Thing STA-Lite:hasTaskingCapability / STA-Lite:URI ?Tasking_URI
}
```

Fig. 10. A SPARQL query example of a door lock in the a-device-as-a-Thing view

Table 2. A SPARQL result of query shown in Fig. 9 and Fig. 10

Door_name/Name	Id	Tasking_URI
Door-Internal-7-1	Fme-gen-796fff29-7661-4bed-bcb9-005dd1824519	http://140.115. 110.69:8080/_sta_ taskingservice_/ STA/v1.0/Thi ng(5)/TaskingCa pabilities

5   Conclusions and Future Work

This research proposes a solution to integrate 3D city model and IoT service standards to support inter-dataset queries between 3D model and IoT resources. To be specific, we design an integration ontology defining the linkage relationships between CityGML and SensorThings API data models. In this case, to IoT resources, sensor observations and tasking capabilities can be analyzed and queried with more spatial information. To 3D model resources, city elements can be extended with dynamic IoT sensing and tasking resources. In addition, multiple views of the IoT Thing are also included in the integration ontology. As shown in the preliminary results, the proposed solution successfully achieves the integration of 3D city model and IoT service standard, which could serve as a necessary component in a smart city infrastructure.

As a future work, to achieve a more comprehensive smart city infrastructure, we plan to include the OGC IndoorGML in the integration ontology by describing relationships between IndoorGML classes with CityGML and SensorThings API classes. Finally, we aim at examining the feasibility of the proposed solution by exploring potential smart city use cases and applications.

References

1. Batty, M.: Big data, smart cities and city planning. Dialogues Hum. Geogr. **3**(3), 274–279 (2013)
2. Sánchez-Corcuera, R., et al.: Smart cities survey: technologies application domains and challenges for the cities of the future. Int. J. Distrib. Sens. Netw. **15**(6), 1550147719853984 (2019)
3. Franklin, C., Hane, P.: An introduction to geographic information systems: linking maps to databases and maps for the rest of us: affordable and fun. Database **15**(2), 12–15 (1992)
4. Zanella, A., Bui, N., Castellani, A., Vangelista, L., Zorzi, M.: Internet of things for smart cities. IEEE Internet Things J. **1**(1), 22–32 (2014)
5. Wang, H.: Sensing information modelling for smart city. In: 2015 IEEE International Conference on Smart City/SocialCom/SustainCom (SmartCity), pp. 40–45 (2015)
6. Sondheim, M., Gardels, K., Buehler, K.: GIS interoperability. geographical. Inform. Syst. **1**, 347–358 (1999)
7. Gröger, G., Kolbe, T.H., Nagel, C., Häfele, K.H.: OGC city geography markup language (CityGML) encoding standard (2012)

8. Liang, S., Huang, C.Y., Khalafbeigi, T.: OGC SensorThings API Part 1: Sensing, Version 1.0. (2016)
9. Berners-Lee, T., Hendler, J., Lassila, O.: The semantic web. Sci. Am. **284**(5), 28–37 (2001)
10. Gruber, T.R.: Toward principles for the design of ontologies used for knowledge sharing? Int. J. Hum Comput Stud. **43**(5–6), 907–928 (1995)
11. Miller, E.: An introduction to the resource description framework. Bull. Am. Soc. Inf. Sci. Technol. **25**(1), 15–19 (1998)
12. Bačić, Ž., Jogun, T., Majić, I.: Integrated sensor systems for smart cities. Tehnički vjesnik **25**(1), 277–284 (2018)
13. Wang, J., Zhao, H., Winter, S.: Integrating sensing, routing and timing for indoor evacuation. Fire Saf. J. **78**, 111–121 (2015)
14. Wang, H., Gluhak, A., Meissner, S., Tafazolli, R.: Integration of BIM and live sensing information to monitor building energy performance. In: The CIB 30th International Conference on Applications of IT in the AEC Industry (2013)
15. Chaturvedi, K., Willenborg, B., Sindram, M., Kolbe, T.H.: Solar potential analysis and integration of the time-dependent simulation results for semantic 3D city models using dynamizers. In: Proceedings of the 12th International 3D GeoInfo Conference 2017, pp. 25–32 (2017)
16. Miorandi, D., Sicari, S., De Pellegrini, F., Chlamtac, I.: Internet of things: vision, applications and research challenges. Ad Hoc Netw. **10**(7), 1497–1516 (2012)
17. OPen Connectivity Foundation (OCF). https://openconnectivity.org/. Accessed 19 Feb 2016
18. Industrial Internet Consortium. https://www.iiconsortium.org/
19. oneM2M-TS-0001, oneM2M Functional Architecture Specification v4.3.0, http://www.one m2m.org/. Accessed 04 Nov 2019
20. Bröring, A., Stasch, C., Echterhoff, J.: OGC® Sensor Observation Service Interface Standard, Open Geospatial Consortium Interface Standard: Wayland, MA, USA (2012) https://www.opengeospatial.org/standards/sos
21. Liang, S., Khalafbeigi, T.: OGC SensorThings API Part 2: Tasking, Version 1.0, Open Geospatial Consortium. OGC. Candidate standard (2018)
22. Cox, S.: Observations and measurements. Open Geospatial Consortium Best Practices Document. Open Geospatial Consortium, 21. (2006)
23. Botts, M., Percivall, G., Reed, C., Davidson, J.: OGC® sensor web enablement: overview and high level architecture. In: Nittel, S., Labrinidis, A., Stefanidis, A., (eds.) GeoSensor Networks, pp. 175–190. Springer, Heidelberg (2006) https://doi.org/10.1007/978-3-540-79996-2_10
24. Pizzot, M., Handl, R., Zurmuehl, M.: Information technology Open data protocol (OData) v4.0 Part 1: Core, OAISI Open Data Protocol (OData) TC (2016). https://www.iso.org/sta ndard/69208.html
25. Vilgertshofer, S., Amann, J., Willenborg, B., Borrmann, A., Kolbe, T.H.: Linking BIM and GIS models in infrastructure by example of IFC and CityGML. Comput. Civil Eng. **2017**, 133–140 (2017)
26. Kim, J.S., Yoo, S.J., Li, K.J.: Integrating IndoorGML and CityGML for indoor space. In: International Symposium on Web and Wireless Geographical Information Systems, pp. 184–196. Springer, Heidelberg (2014). https://doi.org/10.1007/978-3-642-55334-9_12
27. ITU Telecommunication Standardization Sector, Overview of Internet of Things, ITU-T: Geneva, Switzerland (2012). https://www.itu.int/rec/T-REC-Y.2060-201206-I

Location Optimization of Urban Emergency Medical Service Stations: A Hierarchical Multi-objective Model with a New Encoding Method of Genetic Algorithm Solution

Jiajia Song[1] , Xiang Li[1(✉)], and Joseph Mango[1,2]

[1] Key Laboratory of Geographic Information Science (Ministry of Education) and School of Geographic Sciences, East China Normal University, Shanghai, China
xli@geo.ecnu.edu.cn
[2] Department of Transportation and Geotechnical Engineering, University of Dar es Salaam, Dar es Salaam, Tanzania

Abstract. Since patients' lives are closely bound up with emergency medical services, extensive studies to improve the quality of emergency services haves been receiving special attention. This paper presents a novel hierarchical multi-objective optimization model that considers the goal of providing effectiveness equal service for all citizens firstly, reducing the total travel cost of emergency medical service missions and the number of overall stations secondly, retaining as many existing stations as possible lastly to improve both the effectiveness equity and the efficiency of emergency medical service and reduce the financial cost. New methods of chromosome coding, crossover operation and mutation operation for preserving spatial configuration of regional variables in the process of genetic algorithm are developed and used to optimize locations of EMS stations in Shanghai, China. The results demonstrate that better planning of emergency medical service stations whose service area cover all area within 4 km can reduce travel costs by 70% with 13 new built up and 8 existing stations. Due to these promising results, the new encoded methods applied in this study are not only viable but also can be used in other urban areas to improve effectiveness equity of the emergency medical service.

Keywords: Location optimization · Hierarchical multi-objective model · Emergency medical service stations · Effectiveness equity · Binary space mapping chromosome encoding

1 Introduction

Planning of emergency systems has never failed to gain great attention, including security guard operations, fire stations, etc., which are essential in urban systems. The emergency medical services (EMS) often needed in many situations (e.g. after traffic accidents). As injuries continue to grow as a cause of death and disability in someplace in the world [1] and take up about 12% of the disease around the world according to the World Health

S. Di Martino et al. (Eds.): W2GIS 2020, LNCS 12473, pp. 68–82, 2020.
https://doi.org/10.1007/978-3-030-60952-8_7

Organization Report in 2003, many injuries died because of the poor condition of urban emergency medical service system. Thus, it is very critical to establish an effective EMS system that can provide a high quality of service for patients.

Typically, the quality of an emergency medical service system can be evaluated by its response time, i.e., the time between an emergency call and the arrival of an ambulance at the spot. Obviously, time is a crucial factor in improving survival rates of patients [2] and recent findings emphasize the significance of shortening response time to reduce patients' mortality rates. For example, Okeeffe, Nicholl, Turner, & Goodacre [3] presented that a reduction of one-minute of the response time can raise the possibility of survival by 24%. The study by Thompson [4] also pointed out that the most significant predictive factor of the survival of a patient is time. Therefore, the key to improving the effectiveness of an emergency medical service system is to reduce the response time.

The response time that an EMS system can achieve depends on a variety of factors. Since the medical resources such as the number of emergency medical stations and ambulances are limited, then it is impossible to locate ambulances everywhere. Therefore, there must exist a distance between the emergency call places and the nearest emergency medical stations.

In general, the process of medical service consists of following steps [5]: (1) an incoming emergency call, (2) answering the emergency call, (3) dispatching an ambulance, (4) the traveling of the ambulance to the call site, (5) first-aid treatment. All these steps can be affected by several factors such as the traffic situation, population distribution, the potential risks of accidents, the dispatch policy, etc.

So far, considerable amount of researches have been devoted to improve efficiency of the EMS systems. [6–8] contributed to crew pairing, scheduling and familiarity while [9–11] looked at ambulance dispatching or relocation with rules of nearest vehicle, fast-reaching vehicle. In the same line, [12] focused to minimize the cost and optimize the routine of ambulances to avoid unnecessary delays and the waste of resource. Others focused on developing different location models to find optimal placements of EMS stations with different objectives. [13, 14] are about server-to-customer systems take coverage into consideration, trying to solve the maximal covering location problem (MCLP). Other researches including [15, 16] account for the efficiency of the system by solving the p-median problem (PMP) in order to reduce total travel costs.

Recently, the equity and fairness, referring to how well an individual's needs can be satisfied fairly with limited resources have gained great attention. There are mainly three approaches are suggested to seek equity [5]. The first approach involves a function originated in a method that minimize the worst served point proposed by Rawls. The second is to minimize the deviation, variance, such as [17–20]. The third, however, elaborates about suffering and damages caused by late EMS. Little consensus exists on the metrics of equity since it is more a perception varying from person to person, situation to situation, than measurable quantity sometimes. [21–23] used envy to measure the inequity which is difficult to quantify. Since locations of emergency medical service stations in urban regions cannot be changed frequently, it is important to ensure that the worst-off situation in a certain area can get a relatively better service. Thus, it is critical to identify the notion of equity in EMS in a clear and easy way.

In the past few decades, [24, 25] have tried to solve the location problem in a more comprehensive approach that comprised more than one objective in their model. In order to solve the multi-objective model, the Pareto-optimal, which can return a large number of solutions that are Non-dominated by each other, is often used. Thereafter, people must choose one, which can be very blind because every solution is no worse than another. To the best of our knowledge, there is lack of sufficient and clear knowledge for solving these problems. Based on that, this work proposes an easy and practical measurement of equity by using hierarchical multi-objective model that is encoded by new methods for optimizing locations of the EMS stations.

When it comes to the solution to the model, substantially, it is a location-allocation problem that has been extensively studied by many mathematical models and GIS methods, including [26–29].

The most direct and comprehensive approaches for solving the problem of this type include the exhaustive methods which evaluate and compare all possible scenarios. However, in the face of a wide range of complex situations, the methods are difficult to deploy many situations. For example, there are possibilities for selecting 20 positions in cells, which is impossible to enumerate. Later, many mathematical models that belong to the NP-hard problem have been proposed such as the p-Median Problem, p-Center Problem, and the p-Coverage Problem. Algorithms based on graph theory, linear programming and other traditional mathematic methods can also only be used for small-scale problems [30]. It is very difficult to establish mathematical models accurately and reasonably in the face of the actual situation.

In such cases, the intelligent optimization algorithms such as particle swarm algorithm, ant colony algorithm and genetic algorithm have been widely used for complex geographical problems [25, 31] due to their automation and versatility. GA is applied in solving many optimization tasks. However, it does not provide sufficient spatial configuration of the regional variables, hence its ability to optimize location models is uncertain. Therefore, this paper proposes a new method of GA encoding that can preserve the spatial configuration of regional variables. The main contribution of this paper includes the novelty of proposing a new practical notion of equity of EMS and developing a hierarchical multi-objective model in order to 1) minimize the total travel cost while making sure everywhere in study area can get effective EMS, 2) minimize the total number of stations and the number of existing stations that need to be replaced at the same time and 3) propose a new method of binary space mapping chromosome encoding to retain the spatial configuration in GA operators to solve the model. The travel costs in the model are estimated from urban transportation road data instead of using Euclidean distance as applied in other studies [25, 31].

In order to meet the requirements discussed above, this paper will estimate the possibility of having EMS demands in each small area using historical emergency medical service data and then, will establish a hierarchical multi-objective model to optimize the configuration of EMS stations. Finally, it will use the NSGA-II with binary space mapping chromosome encoding to solve the model. All the approaches and methods will use the emergency medical incident data collected from 2012–2017 in Min Hang District, Shanghai, China. The paper also offers findings and suggestions for future extensions.

2   Methods

According to the discussion above, both equity and efficiency of the EMS system are account to optimize the spatial distribution of EMS stations. Therefore, this section introduces a hierarchical multi-objective location model for locating EMS stations that balance the equity and efficiency and improve GA with a newly proposed space binary mapping encoding method.

Extensive literature referring to a server-to-customer system like the EMS system has discussed the measure to evaluate the equity and efficiency of a system [32]. The measure of efficiency which has been widely used is the total travel cost. In this paper, we used the weighted travel cost overall the study area to measure the efficiency. Further, to minimize financial and time costs, we tried to establish few stations and retain many old stations which are important in measuring efficiency.

The aim of EMS is to ensure that the survival rate of patients. With the limited number of EMS stations (EMS stations cannot be found everywhere) and the relatively fixed locations of urban EMS stations (Locations of EMS stations cannot change frequently), the notion of equity can just be based on the effectiveness. Specifically, if an EMS demand call can be solved within a life support time, this EMS mission can be considered that it is effective. Thus, the notion of equity we proposed in this paper is that, all people no matter where they are can get the EMS within the time support time. And we call it effectiveness equity. According to the National Fire Protection Association (NFPA) 2010, an Advanced Life Support (ALS) response is within eight minutes. In this paper, consider the minimum average speed of an ambulance is to be 500 m per minute, we set the predetermined travel cost as 4000 m.

With this in mind, we proposed four objective functions for improving efficiency and effectiveness equity using as few stations totally as possible while conserving as many old stations as possible:

- The travel cost of all missions should be limited within a predetermined value
- Minimize the total weighted travel cost of EMS demands in the study area
- Minimize the number of stations established in the study area
- Maximize the number of existing stations

And the parameters and formulations of the model are as follows:

I: the set of demand areas;
J: the set of potential EMS station locations;
i: index of demand areas;
j: index of potential EMS locations;
w_i: estimated the possibility of emergency demand in area i, calculated by the portion of emergency calls happen in the area i;
d_{ij}: travel cost from station j to an emergency call demand area, usually the distance or the travel time;
C: a set of points refers to existing stations
T: the predetermined maximum travel cost within which a mission is effective;

$$Y_j = \begin{cases} 1, & \textit{if an EMS station is located at area } j \\ 0, & \textit{otherwise} \end{cases}$$

$$E_{ij} = \begin{cases} 1, & \textit{if } d_{ij} \leq T \\ 0, & \textit{if } d_{ij} > T \end{cases}$$

$$\text{Max } j \in Ji \in Idij < T \tag{1}$$

$$\text{Minimize } j \in Ji \in Iwidij \tag{2}$$

$$\text{Minimize } j \in JYj \tag{3}$$

$$\text{Maximize } j \in CYj \tag{4}$$

Subject to:

$$Y_j = \{0, 1\} \forall j \in J \tag{5}$$

Objective (1) aims to ensure all grids covered by effective EMS, and objective (2) focuses on minimizing the total weighted travel cost of all emergency calls. Objective (3) minimizes total number of EMS stations established in the study area while objective (4) means when the total number of EMS stations is the same, is to try to preserve as many existing stations as possible. Constraints (5) require that the total number of EMS stations to be less than or equal to the expected number of stations. All the EMS missions will be assigned to the nearest station.

Noting that it is a multi-objective model, it desires a Pareto-optimal solution. Different from non-Pareto solutions that cannot simultaneously optimize all objectives, Pareto-optimal solution, however, makes sure that one objective will not be improved with degrading the other objective [33]. A myriad of Pareto-optimal solutions has been proposed to solve this model, like the Non-dominated Sorting Genetic Algorithm (NSGA) [34]. But it should be noted that with the increasing size of J, as this paper trying to search for an optimal solution in the whole area, the computational complexity will increase even faster. On this account, we selected Non-dominated Sorting Genetic Algorithm-II (NSGA-II) as the basic algorithm. It is said that NSGA-II performs better than other elitist multi-objective evolutionary algorithms in the regard of converging near the Pareto-optimal front and keep the diversity of solutions obtained at the same time and can run much faster with the maximum complexity of [34].

However, the traditional methods of chromosome encoding, including float code, gray code, binary code, etc., cannot perfectly represent spatial variables whose spatial configuration will be damaged in the process of encoding. Worse off, without the spatial configuration, genetic operators in the following steps cannot serve as a driving force to generate a better generation. To address this problem, we proposed a new method of chromosome encoding– binary space mapping code. This method of encoding can fully reappear and retain the spatial configuration of study objects. The details of coding, GA crossover, and GA mutation are as follows:

- Coding and forming initial populations

  $P = \{P_1, P_2, \dots P_n\}$ is regarded as in initial populations, and n is the number of individuals in a population. P_u represents a matrix of $Y = \begin{pmatrix} y_{11} & \cdots & y_{1m} \\ \vdots & \ddots & \vdots \\ y_{n1} & \cdots & y_{nm} \end{pmatrix}$, $y_{ij} \in \{0, 1\}$, n, m refers to the number of grids in a row and the number of grids in a column in the study area respectively. If the value of y_{ij} is 1, a station is locating at the grid (i, j), otherwise, the value of y_{ij} is 0. If the locations of potential stations are limited to a few areas (grids), the y of other grids can be fixed to 0.

- GA crossover

  To preserve some parts of the spatial configuration of good individuals, we cut the space mapping chromosome using two approaches randomly: cut it vertically or horizontally. It should be noted that the selection of breakpoints and the cutting approach must work together to ensure that the crossover operator is valid, which means there must be something different between the parent generation and the offspring. For example, if parent individuals aC_1 and bC_2 are cut in point c horizontally, the operation is invalid because there is no difference between two generations. A wide range of cutting approaches can be applied, such as stepped cutting, to meet requirements of different objectives. One can even cut the space mapping chromosome into several parts and exchange parts among several individuals (Fig. 1). There are two parent individuals P_1, P_2, and the filled grid means there is a station established there while the white one means there is no station. The yellow grids stem from P_1 and the black one stems from P_2. If cut two parent individuals in point a vertically, in the same way, two children individuals aC_1 and aC_2 will be generated. If cut individuals in another breakpoint b horizontally, two children individuals bC_1 and bC_2 will be generated. If parent individuals aC_1 and bC_2 are cut in point c horizontally, the children individuals will be the same as parent individuals aC_1 and bC_2.

- GA mutation

  Like the crossover operator, the mutation operator is very flexible, too. We can set different mutation rules to meet different constraints. In this paper, however, in terms of model objectives and model settings: we seek the optimal solution that has least stations whose service area can cover the whole region, it is good to try new locations of stations and remove some bad locations. On account of this, we allowed the mutation to happen in all grids, and in each time, we selected a small percentage of the grid to mutate from 1 to 0 or from 0 to 1 randomly.

- Constraints

  When the model generates a new population, we have to make sure that it subjects to constraints (5). In the process of sorting, we adjusted its principle of dominance to fit our hierarchical multi-objective model. Objective (1) has the top priority, and objective (2), (3) share the same rank while objective (4) is secondary to objective (3) which means if and only if two individuals have the same value of objective (3), objective (4) works. The difference has shown as below:

  The traditional definition of dominance:

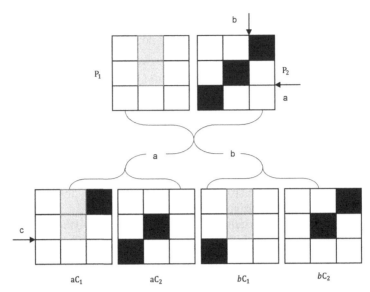

Fig. 1. The process of GA crossover of binary space mapping coding individuals (Color figure online)

x_1 dominates x_2, if the solution a_1 is no worse than a_2 in overall objectives and the solution a_1 is strictly better than a_2 in at least one objective.

Definition of dominance in our hierarchical multi-objective model:

x_1 dominates x_2, if the solution a_1 is not worse than a_2 in overall objectives in the first rank, if objectives in the first rank are the same, solution a_1 is no worse than a_2 in overall objectives in the second rank and so on and the solution a_1 is strictly better than a_2 in at least one objective among all objectives.

The overall procedures are presented in Fig. 2. Firstly, setting the population size, the largest iteration I and generation initial population P_0, and let the index of iteration i = 0. Secondly, execute the genetic operations, namely crossover operation, mutation operation and obtain the child generation Q_t. And then integrate the parent generation P_t and the child generation Q_t as R_t. After that, apply the sorting and density compare process to R_t to select elites as the new generation P_{t+1} and then redo the genetic operations until the total iteration reaches I.

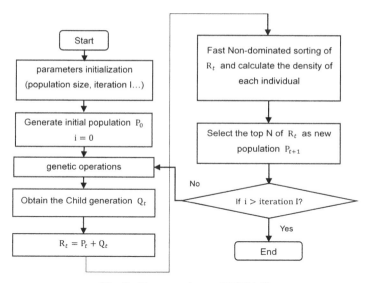

Fig. 2. The procedures of NSGA-II

3 Case Study

3.1 Study Area and Data

This section presents the information about case study area where this research is conducted. The implementation of our research methods has used data from Min Hang EMS provider found in the city of Shanghai, China. According to the China Census Bureau in 2017, Shanghai estimated a population of more 2.5 million in the area of 372.56 km². The EMS in the city has 29 ambulances and 12 EMS stations. From the period of January 2012 to December 2017, the city recorded a total of 205414 emergency medical incidents. Each record has an address to indicate the calling position and the call time from Medical Emergency Center.

In practical terms, we used urban transportation roads network data obtained from Geographic Data Sharing Infrastructure, College of Urban and Environmental Science, Peking University (http://geodata.pku.edu.cn) to estimate the travel cost from each potential station to each emergency call site.

We divided the study area into regular grids with a size of 100 m × 100 m. The whole study area has been covered by 3184 grids and the possibility of having EMS demands in each grid was estimated from the historical EMS data as specified in the years above. Each district in the study area has one station. However, some stations are not located in the areas with higher EMS demands. Additionally, only 1045 demand areas have the EMS stations within a service standard of 4 km, leaving 2139 (more than 67% of the whole region) uncovered, which contains more than 32% demands of EMS. The sum of the weighted distance from the demand points to their nearest station is about 13.27 km (Fig. 3). Geographically, the case study area falls between 30.975°N to 31.264°N latitudes and from 121.235°E to 121.572°E longitudes by the WGS 1984 coordinate reference system. And the service area is based on urban transportation road network data.

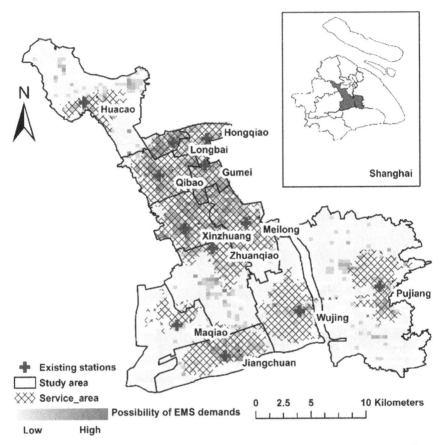

Fig. 3. Study area, existing station, service area and the possibility of EMS demands

3.2 Optimal Locations of EMS Station

To ensure that we are able to find a solution that can meet the requirement of objective (1), we took all the locations that satisfy the conditions of establishing a station there as potential locations of stations including positions of existing stations (Fig. 4(a)).

Using the methods discussed above, we initialized 50 individuals in the first generation and run 100 iterations. There are 60 potential locations of EMS stations distributed in the study area, among which there are 12 existing EMS stations. After 100 iterations, we got an optimal solution with 13 new stations and 8 existing stations, and anywhere in the study area can get EMS within a travel distance of 4000 m (Fig. 4(b)).

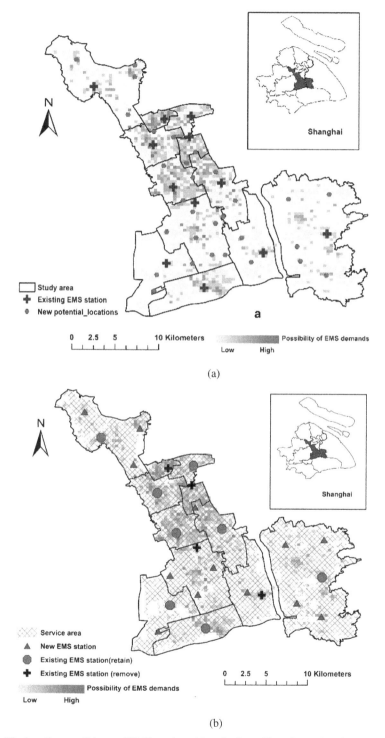

Fig. 4. The location candidates of EMS stations (a) and selected locations of stations (b) with the service area

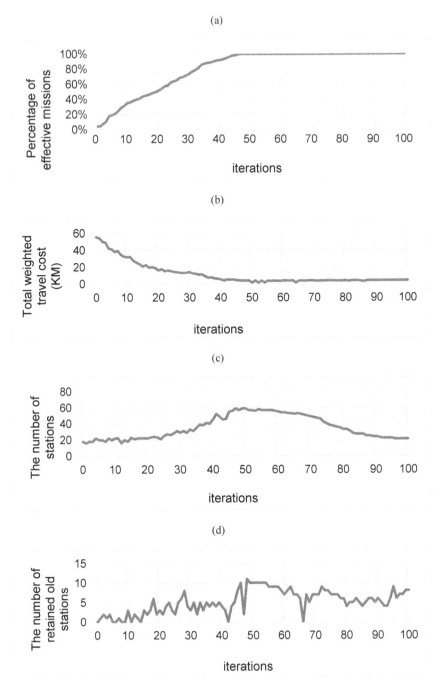

Fig. 5. The performance of best individuals among the Pareto-front individuals (a) the performance of objective (1): the percentage of effective missions (within the distance of 4 km); (b) the performance of objective (2): the total weighted distance; (c) the performance of objective (3): the number of stations should be established; (d) the performance of objective (4): the number of retained existing stations.

There are 48 new location candidates of EMS stations and 12 existing stations in (a). According to the nature of our study, all these stations serve as the location candidates in the optimization model. After running the model, it showed the demand of establishing new 13 EMS stations and retaining 8 of 12 existing stations while the other 4 stations should be relocated.

In the process of running the genetic algorithm, the footprints of the best individual in each iteration have been recorded. It should be noted that in this process, we chose an individual in the Pareto-front that has the least station number as the best individual. As indicating in Fig. 5(a), the objective (1) with the top priority has been optimized first in the first 45 iterations until it reached a maximum of 100%. Despite of having some small fluctuations, the value of objective (2) is decreasing gradually (Fig. 5(b)). This trend is attributed by the fact that, as the percentage of missions increases for stations found within 4000 m, the total weighted distance drops as hypothesized. The final value of the objective (2) is 3.981 km. When the value of objective (1) peaks in the iteration of about 45, the number of stations reaches its top value. This is because as more stations established in the study area, then the distance between the EMS demand area and the nearest station becomes shorter. Afterward, the second stage of optimizing follows. The number of stations begins to decrease, ending up at the number of 21 (Fig. 5(c)). Different from the other three values of objectives, the number of retained old stations seems to fluctuate greatly (Fig. 5(d)). However, this result is reasonable since it has the lowest rank of being optimized, and every time it keeps increasing while the values of other objectives remained the same.

The final solution of EMS station planning with 13 new stations and 8 existing stations, achieves a 70% improvement of total weighted distance (TWD) of 3.981 km compared with the existing plan of EMS station distribution whose TWD is 13.27 km and its service area covers all the places in the study area.

4   Discussion and Conclusion

Pre-historical aid is an important part of emergency medicine whose significance lies in saving patients' lives and winning time and treatment conditions for hospital treatment. Thus, the ability to provide the first line of response in a very short time for urban EMS is fundamental to the safety of citizens. This paper has proposed a multi-objective optimization model to explore the optimal spatial configuration of EMS stations, accounting for offering equal service for all citizens, reducing the total travel cost of EMS missions and retaining as many existing stations as possible to improve both the effectiveness equity and the efficiency of EMS and reduce the financial cost by considering potential requirements.

Admittedly, there is a wide range of relevant literature on planning customer service stations. However, in this paper, we proposed a hierarchical method to optimize several objectives with different priorities at a time. We opted this after realizing that many classical methods applied perfectly in other fields of study but cannot fully work at geography because of the particularity of regionalized variables. Most of the studies have used classical statistics which cannot deal with spatial relativeness of regionalized variables. The GA is without spatial information and configuration of regionalized variables will be damaged while encoding. Worse off, the GA operators lose the meaning of generating better offspring. Therefore, we proposed the new encoding method called, binary space mapping chromosome encoding and introduce some basic approaches about its crossover and mutation operators for preserving the configurations and other spatial information through encoding and act as a driving force in GA operating. Taken altogether, this method is very flexible and can be easily developed for use in other researches. Importantly, we estimated travel cost based on the transportation road data which is closer to the reality rather than using Euclidean distance.

The proposed methods and model for optimizing EMS station configuration are applied in shanghai, china. The empirical findings have crucial implications for the policy of local governments and the medical system. For example, the current planning of EMS stations is based on administrative divisions (e.g. one station per county), which is not reasonable: a wide range of areas is not covered by effective EMS as Fig. 3 indicated. The results from this study suggest that more EMS stations should be established in Pujiang county, Maqiao county, Zhuanqiao county, and Huacao county. The solution shown in Fig. 4(b) satisfies the EMS demand of citizens with the least number of stations and the least travel cost in a long term, thus it can be applied to improve the quality of the EMS system greatly.

In future, this work can be extended in several ways. As discussed in the introduction, [35–37] regard whether a position will have an EMS demand in a specific moment or not as a random event. However, in this paper, we expected that the possibility of EMS demand in a relatively wide range of area could be predicted from historical EMS data since [25] used the density distribution of urban population or historical data to estimate the possibility. Taking those concerns into consideration, a study should be done to explore if there are some rules that exist for the temporal and spatial patterns of EMS demands in the urban area. Moreover, we have only considered spatial planning of EMS stations in an urban area without dealing with their service capacity and the management of medical resources. Thus, more objectives and constraints can be added in the model to obtain more meaningful results. Furthermore, due to the limitation of historical or real-time traffic condition data of the study area, we used the travel cost of 4 km as a given measurement to evaluate the effectiveness of an EMS mission. However, it is the time rather than distance that determines an EMS mission successful or not. Hence, the traffic condition data will be required to estimate the travel cost in further study

Acknowledgments. We are indebted to anonymous reviewers for insightful observations and suggestions that helped to improve our paper. This work is partially supported by the projects funded by the National Nature Science Foundation of China (grant numbers: 41771410 and 41401173).

References

1. Karbakhsh, M., Zandi, N., Rouzrokh, M., Zarei, M.: Injury epidemiology in Kermanshah: The National Trauma Project in Islamic Republic of Iran. Eastern Mediterranean health journal = La revue de santé de la Méditerranée orientale = al-Majallah al-ṣiḥḥīyah li-sharq al-mutawassiṭ **15**, 57–64 (2009)
2. Wilde, E.: Do emergency medical system response times matter for health outcomes? Health Econ. **22**, 790–806 (2013)

3. O'Keeffe, C., Nicholl, J., Turner, J., Goodacre, S.: Role of ambulance response times in the survival of patients with out-of-hospital cardiac arrest. Emerg. Med. J.: EMJ **28**, 703–706 (2011)
4. Gonzalez, R., Cummings, G., Phelan, H., Mulekar, M., Rodning, C.: Does increased emergency medical services prehospital time affect patient mortality in rural motor vehicle crash? A statewide analysis. Am. J. Surg. **197**, 30–34 (2008)
5. Bélanger, V., Ruiz, A., Soriano, P.: Recent optimization models and trends in location, relocation, and dispatching of emergency medical vehicles. Eur. J. Oper. Res. **272**, 1–23 (2018)
6. Eschmann, N., Pirrallo, R., Aufderheide, T., Lerner, E.: The association between emergency medical services staffing patterns and out-of-hospital cardiac arrest survival. Prehosp. Emerg. Care: Off. J. Natl. Assoc. EMS Physicians Natl. Assoc. State EMS Dir. **14**, 71–77 (2009)
7. Lujak, M., Billhardt, H.: A distributed algorithm for dynamic break scheduling in emergency service fleets. In: An, B., Bazzan, A., Leite, J., Villata, S., van der Torre, L. (eds.) PRIMA 2017. LNCS (LNAI), vol. 10621, pp. 477–485. Springer, Cham (2017). https://doi.org/10.1007/978-3-319-69131-2_30
8. Weaver, M., Patterson, P., Fabio, A., Moore, C., Freiberg, M., Songer, T.: The association between weekly work hours, crew familiarity, and occupational injury and illness in emergency medical services workers. Am. J. Ind. Med. **58**, 1270–1277 (2015)
9. Bandara, D., Mayorga, M., Albert, L.: Optimal dispatching strategies for emergency vehicles to increase patient survivability. Int. J. Oper. Res. **15**, 195–214 (2012)
10. Carter, G., Chaiken, J., Ignall, E.: Response area for two emergency units. Oper. Res. **20**, 571–594 (1972)
11. Schmid, V.: Solving the dynamic ambulance relocation and dispatching problem using approximate dynamic programming. Eur. J. Oper. Res. **219**, 611–621 (2012)
12. Tlili, T., Harzi, M., Krichen, S.: Swarm-based approach for solving the ambulance routing problem. Proc. Comput. Sci. **112**, 350–357 (2017)
13. Karasakal, O., Karasakal, E.: A maximal covering location model in the presence of partial coverage. Comput. Oper. Res. **31**, 1515–1526 (2004)
14. Atta, S., Sinha Mahapatra, P.R., Mukhopadhyay, A.: Solving maximal covering location problem using genetic algorithm with local refinement. Soft. Comput. **22**(12), 3891–3906 (2017). https://doi.org/10.1007/s00500-017-2598-3
15. Reilly, J., Mirchandani, P.: Development and application of a fire station placement model. Fire Technol. **21**, 181–198 (1985). https://doi.org/10.1007/BF01039973
16. Richard, D., Beguin, H., Peeters, D.: The location of fire stations in a rural environment: a case study. Environ. Plan. A **22**, 39–52 (1990)
17. Berman, O., Kaplan, E.: Equity maximizing facility location schemes. Transp. Sci. **24**, 137–144 (1990)
18. Brill, E., Liebman, J., ReVelle, C.: Equity measures for exploring water quality management alternatives. Water Resour. Res. **12**, 845–851 (1976)
19. Erkut, E., Neuman, S.: A multiobjective model for locating undesirable facilities. Ann. Oper. Res. **40**, 209–227 (1993). https://doi.org/10.1007/BF02060478
20. Kincaid, R., Maimon, O.: Locating a point of minimum variance on triangular graphs. Transp. Sci. **23**, 216–219 (1989)
21. Espejo, I., Marín, A., Puerto, J., Rodríguez-Chía, A.: A comparison of formulations and solution methods for the minimum-envy location problem. Comput. Oper. Res. - CoR **36**, 1966–1981 (2009)
22. Chanta, S., Mayorga, M., Kurz, M., Albert, L.: The minimum p-envy location problem: a new model for equitable distribution of emergency resources. IIE Trans. Healthc. Syst. Eng. **1**, 101–115 (2011)

23. Chanta, S., Mayorga, M., Albert, L.: The minimum p-envy location problem with requirement on minimum survival rate. Comput. Ind. Eng. **74**, 228–239 (2014)
24. Araz, C., Selim, H., Ozkarahan, I.: A fuzzy multi-objective covering-based vehicle location model for emergency services. Comput. Oper. Res. **34**, 705–726 (2007)
25. Yao, J., Zhang, X.: Location optimization of fire stations: trade-off between accessibility and service coverage. In: International Conference on GIScience Short Paper Proceedings, vol. 1 (2016)
26. Esmaelian, M., Tavana, M., Santos Arteaga, F., Mohammadi, S.: A multicriteria spatial decision support system for solving emergency service station location problems. Int. J. Geogr. Inf. Sci. **29**, 1187–1213 (2015)
27. Flamand, T., Ghoniem, A., Haouari, M., Maddah, B.: Integrated assortment planning and store-wide shelf space allocation: an optimization-based approach. Omega **81**, 134–149 (2017)
28. He, D., Jia, R.: Cloud model-based Artificial Bee Colony Algorithm's application in the logistics location problem. In: 2012 International Conference on Information Management, Innovation Management and Industrial Engineering (ICIII), pp. 256–259. IEEE Press (2012)
29. Quintero-Araujo, C., Gruler, A., Juan, A., Faulin, J.: Using horizontal cooperation concepts in integrated routing and facility-location decisions. Int. Trans. Oper. Res. **26**, 551–576 (2017)
30. Colome Perales, R., Lourenço, H., Serra, D.: A new chance-constrained maximum capture location problem. Ann. Oper. Res. **122**, 121–139 (2003). https://doi.org/10.1023/A:102619 4423072
31. Albert, L., Mayorga, M.: A dispatching model for server-to-customer systems that balances efficiency and equity. Manuf. Serv. Oper. Manag. **15**, 205–220 (2013)
32. Enayati, S., Mayorga, M., Toro-Díaz, H., Albert, L.: Identifying trade-offs in equity and efficiency for simultaneously optimizing location and multipriority dispatch of ambulances. Int. Trans. Oper. Res. **26**, 415–438 (2018)
33. Current, J., Revelle, C., Cohon, J.: An interactive approach to identify the best compromise solution for two objective shortest path problems. Comput. Oper. Res. **17**, 187–198 (1990)
34. Tran, K.: Elitist Non-Dominated Sorting GA-II (NSGA-II) as a parameter-less multi-objective genetic algorithm. In: Proceedings, pp. 359–367. IEEE, FL, USA (2005)
35. van Barneveld, T.C., Bhulai, S., van der Mei, R.D.: A dynamic ambulance management model for rural areas. Health Care Manag. Sci. **20**(2), 165–186 (2015). https://doi.org/10.1007/s10 729-015-9341-3
36. Shariat, A., Babaei, M., Moadi, S., Amiripour, S.: Linear upper-bound unavailability set covering models for locating ambulances: application to Tehran rural roads. Eur. J. Oper. Res. **221**, 263–272 (2012)
37. Su, Q., Luo, Q., Huang, S.: Cost-effective analyses for emergency medical services deployment: a case study in Shanghai. Int. J. Prod. Econ. **163**, 112–123 (2015)

Registration of Multi-scan Forest Terrestrial Laser Scanning Data Integrated with Smartphone

Maolin Chen$^{(\boxtimes)}$ (iD), Feifei Tang, and Jianping Pan

School of Civil Engineering, Chongqing Jiaotong University, No. 66, Xuefu Avenue, Nan'an District, Chongqing, China
maolinchen@whu.edu.cn

Abstract. Terrestrial laser scanning (TLS) is an important technique to obtain the side view data under the forest canopy and the registration of TLS data captured from different positions is an important step to obtain a complete forest dataset. As commonly used TLS data registration methods are not suitable for the forest scene, this paper presents a registration method of forest TLS data based on orienting and positioning data of smartphone. The coarse transformation parameters between two scans are calculated based on the initial scanner direction and position of each scan, and then used as the input of the Iterative Closest Point (ICP) algorithm for stem position points to get the fine transformation parameters. The experimental results show that this method can achieve accurate alignment of the stem points captured from different sides in case of no artificial targets and poor intervisibility between different scans.

Keywords: Terrestrial laser scanning · Forestry · Registration

1 Introduction

1.1 A Subsection Sample

Forest is an important part of the ecosystem and obtaining single tree parameters in the forest is an important task in many areas, e.g., geological disaster prevention and forest investigation [1]. Traditional forest investigation method is commonly based on manual field measurement, which is time-consuming and difficult to reflect the dynamic changes under large-scale forest environment [2, 3]. Terrestrial laser scanning (TLS) can obtain dense 3D points on the surface of the target without contact and damage, and has been used in many forest investigation work, e.g., tree stem extraction and reconstruction [4], vegetation density estimation [5], biomass estimation [6], DBH and tree height measurement [7]. Multiple scans are commonly needed to cover the whole scanning scene in TLS and thus the scanning data from different scans needs to be converted to the same coordinate system, which is called registration.

Iterative Closest Point (ICP) is the most widely used registration method [8, 9], which iteratively searches adjacent points and calculate transformation parameters between two

© Springer Nature Switzerland AG 2020
S. Di Martino et al. (Eds.): W2GIS 2020, LNCS 12473, pp. 83–89, 2020.
https://doi.org/10.1007/978-3-030-60952-8_8

groups of TLS data until the convergence condition is satisfied. The drawback of ICP is that the iteration will probably converge to a local minimum without a suitable initial parameter. Thus many studies about coarse registration have been made to provide initial transformation parameters for ICP, based on, e.g., reflectance [10], geometric features [11] and semantic information [12]. However, current registration methods mainly face to urban area, commonly containing lots of distinctive geometric features, and may be inapplicable under forest condition. A common method for TLS data registration in forest environment is placing artificial targets as connection points/objects among different scans. But this method brings extra work and is limited by the visibility especially in dense forest. To address this problem, zhang et al. placed a reflector at the position of each scan to recover the translating relationship and utilized the internal compass of the scanner to recover the rotation relationship between two adjacent scans [13]. This method can achieve high scanning efficiency and registration accuracy, but places orientating requirement to the scanner and is still limited by the visibility between different scan positions. In order to avoid the utilization of artificial targets, Kelbe et al. constructed stem position triples to calculate transformation parameters [14]. Similarly, the stem positions were also utilized in the studies of Henning et al. [15] and Liang et al. [16] to achieve non-reflector registration. Those methods get rid of the target dependence, but the registration may fail when there are not enough overlapping stem positions.

It can be found that most of the previous studies on registration focus on urban data while the characteristic of forest scene require different train of thought for registration. The artificial targets and stem positions are the main connection objects used in forest TLS data registration, which are affected by visibility and overlapping rate. To address those problems, a forest TLS data registration method integrated with smart phone is proposed in this article. The coarse transformation parameters are calculated based on the initial orientation and position of the scanner captured by the smartphone without the requirement of artificial targets. Then the coarse parameters are refined using the stem positions to obtain the final transformation.

2   Registration

2.1   Coarse Registration

Our research follows the normal assumption that the scanner is set roughly horizontally during scanning [17], which is achieved by the smart phone gyroscope here. The geodetic coordinates (B, l) of the scanner position and the clockwise angle α between north and the initial orientation can be obtained by smartphone, as shown in Fig. 1 (a) and (b). Thus, the horizontal translation parameters between two scans can be calculated by

$$
\begin{aligned}
D_x &= X_{ref} - X_{sou} \\
D_y &= Y_{ref} - Y_{sou}
\end{aligned}
\tag{1}
$$

where (X_{ref}, Y_{sou}) and (X_{tar}, Y_{sou}) are the Gauss projection coordinates of the reference and source scans respectively, which are converted from the corresponding geodetic coordinates (B_{ef}, L_{ref}) and (B_{sou}, L_{sou}). In this research, the source data is registered to the reference scan. Then the horizontal spatial relationship between the two scans can be constructed by

$$R(-(\theta + \alpha_{ref}))\begin{bmatrix} x_{ref} \\ y_{ref} \end{bmatrix} = R(-(\theta + \alpha_{sou}))\begin{bmatrix} x_{sou} \\ y_{sou} \end{bmatrix} + \begin{bmatrix} D_x \\ -D_y \end{bmatrix} \quad (2)$$

where α_{ref} and α_{sou} are the clockwise angles between north and the initial orientation of the reference and source scans respectively, θ is the clockwise angle between the initial orientation of the scanner and the x-axis of the scanning data and R is the rotation matrix that rotates the x-axis of the two scans to the north direction counterclockwise, as shown in Fig. 1(c). The θ value is a constant, which is 0 for most types of scanners. Notice that, the translation along y-axis should be the opposite number of D_y as the projection coordinates are under the left-handed coordinate system. The horizontal transformation parameters can be obtained based on formula (2)

$$\begin{bmatrix} x_{ref} \\ y_{ref} \end{bmatrix} = R(\alpha_r)\begin{bmatrix} x_{sou} \\ y_{sou} \end{bmatrix} + \begin{bmatrix} d_x \\ d_y \end{bmatrix} \quad (3)$$

with rotation angles and translation parameters are

$$\alpha_r = \alpha_{ref} - \alpha_{sou}, \quad d_x = D_x \cos \alpha_r + D_y \sin \alpha_r$$
$$d_y = D_x \sin \alpha_r - D_y \cos \alpha_r \quad (4)$$

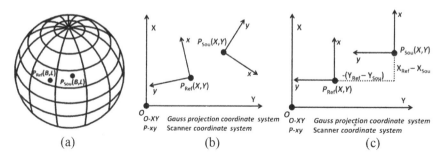

(a)   (b)   (c)

Fig. 1. Process of coordinate transformation

2.2 Fine Registration

Due to canopy occlusion and limitation of smartphone, there are probably errors in the positioning and orienting data, resulting in the registration error. Thus the coarse transformation parameters need further refinement.

The translation parameters in formula (2) can be simplified to the distance between the two scan positions, by rotating the line connecting the two scan positions around the reference scan position horizontally to the parallel direction of the y-axis of the projection coordinate system, as shown in Fig. 2. Thus the horizontal relationship between the reference and source scans are converted to

$$R(\alpha_{ref})\begin{bmatrix} x_{ref} \\ y_{ref} \end{bmatrix} = R(\alpha_{sou})\begin{bmatrix} x_{sou} \\ y_{sou} \end{bmatrix} + \begin{bmatrix} D_{r-s} \\ 0 \end{bmatrix}, \quad \alpha_{ref} = -(\theta + \alpha_{ref}) + \beta_r, \quad \alpha_{sou} = -(\theta + \alpha_{sou}) + \beta_r \tag{5}$$

where D_{r-s} is the distance between the two scan positions and β_r is the rotation angle calculated by

$$\beta_r = \begin{cases} arc\ \cos(-D_y/D_{r-s}), & D_x \geq 0 \\ -arc\ \cos(-D_y/D_{r-s}), & D_x < 0 \end{cases} \tag{6}$$

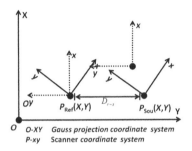

Fig. 2. The orientation of the two scanners after rotating the line connecting the two scanner positions around the reference scan position horizontally to the parallel direction of the y-axis of the projection coordinate system

The stem positions are extracted for each scan data using the method in [4]. After the coarse registration, the corresponding stem positions from the two scans are close to each other. Two position points are recognized as corresponding points if the distance between them is within 0.2 m. In formula (5), the horizontal transformation are determined by the parameter set $(\alpha_{ref}, \alpha_{sou}, D_{r-s})$, which can be generated from the orienting and positioning data of the smartphone. The corresponding registration error can be measured by the root mean square error (RMSD)

$$RMSD = \sqrt{1/n \sum_{i=1}^{n} (p_i - q_i)^2}, \quad p_i \in P, \ q_i \in Q \tag{7}$$

where Q and P are the corresponding point sets from the reference and source scans, p_i and q_i represent the 2D projection of the corresponding points on x-o-y plane.

The most suitable horizontal transformation parameters can be found by searching for the minimum RMSD among different parameter sets

$$(\alpha_{ref}^*, \alpha_{sou}^*, D_{r-s}^*) = arg\ min\ RMSD(\alpha_{ref}, \alpha_{sou}, D),$$
$$\alpha_{ref} \in [\alpha_{ref}^{base} - 10°, \alpha_{ref}^{base} + 10°], \ \kappa_{ref} \in [\alpha_{sou}^{base} - 10°, \alpha_{sou}^{base} + 10°], \ D \in [D_{r-s}^{base} - 10, D_{r-s}^{base} + 10] \tag{8}$$

where ($\alpha base\ ref$, $\alpha base\ sou$, $Dbase\ r - s$) is the initial parameter set calculated from the smartphone data by formula (5) and (α^*_{ref}, α^*_{sou}, D^*_{r-s}) is the parameter set corresponding to the minimum RMSD. The interval of the search range for the angle and distance are set as $1°$ and 1 m respectively. Thus the transformation parameters in formula (2) can be obtained by

$$\begin{bmatrix} x_{ref} \\ y_{ref} \end{bmatrix} = R(\alpha^*_{sou} - \alpha^*_{ref}) \begin{bmatrix} x_{sou} \\ y_{sou} \end{bmatrix} + R(-\alpha^*_{ref}) \begin{bmatrix} D^*_{r-s} \\ 0 \end{bmatrix} \qquad (9)$$

After the stem positions have been aligned in the x-o-y plane, the corresponding 3D stem position points are used as the input of the ICP algorithm to obtain the final transformation parameters.

3   Experimental Results

Two scans of forest data captured by Reigl-VZ400 are used for experiments, with angular resolution of $0.04°$. The point numbers of the reference and source scans are 25986070 and 25602951. The scan positions, terrain model and stem extraction results are shown in Fig. 3. The two scan positions are invisible from each other with height difference of about 8 m and horizontal distance of about 27 m.

Fig. 3. The scanner positions of the reference and source scans in the scene, which are invisible from each other. The horizontal and vertical distances are about 27 m and 8 m

As there are a few artificial objects in the scene, such as footpath, stair, ditch and banner hanging on the tree, reference transformation parameters can be obtained by pick corresponding points on them as the input of ICP. Then the registration result of our method is compared with the result of reference transformation, as shown in Table 1. It can be seen that the proposed method achieves the registration of forest scanning data in case of invisibility between two scan positions and no artificial targets. Although the registration error is higher than that of the reference, Figs. 4 and 5 show that corresponding stem points captured from different sides are well aligned, which is consistent with the smaller horizontal error (RMSD of x and y) in Table 1.

Table 1. Registration error in different steps

	RMSD/m	RMSD of x and y/m	RMSD of z/m
Coarse Registration	0.127	0.079	0.209
Fine registration	0.056	0.038	0.176
Reference	0.026	0.020	0.050

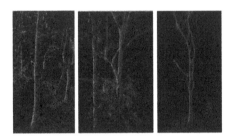

Fig. 4. Side view of the registration results of corresponding stems.

Fig. 5. Nadir view of the registration results of corresponding stem points from two scans.

It is also shown in Table 1 that the vertical error (RMSD of z) is not well refined in the fine registration step. The probable reason is that the ground data is missing at the position of corresponding stem and can only be estimated by nearby ground points. Thus it causes larger error of the z coordinate of the stem position, as the x and y coordinates can be accurately calculated based on the center of the stem points at the bottom. Additionally, the ground may be too undulating to obtain similar z coordinate from different sides of a stem.

4 Conclusion

In this paper, we have presented a registration method for forest terrestrial laser scanning data based on the smart phone orienting and positioning data. The experimental results show that the proposed method can achieve centimeter-level registration of the stem points captured from different sides horizontally, providing reliable trunk information for the post-processing of forest TLS data. A further improvement will focus on the improvement of the registration accuracy on vertical direction.

Acknowledgements. This work was supported in part by National Natural Science Foundation of China under grant no. 41801394, in part by Chongqing Natural Science Foundation under grant no. cstc2019jcyj-msxmX0370 and cstc2018jcyjAX0515 and in part by the Science and Technology Research Program of Chongqing Municipal Education Commission under grant no. KJQN201900729 and KJQN201900728.

References

1. Côté, J.F., Fournier, R.A., Egli, R.: An architectural model of trees to estimate forest structural attributes using terrestrial LiDAR. Environ. Model Softw. **26**(6), 761–777 (2011)
2. La, H.P., Eo, Y.D., Chang, A., et al.: Extraction of individual tree crown using hyperspectral image and LiDAR data. KSCE J. Civ. Eng. **19**(4), 1078–1087 (2015)
3. Liu, F., Tan, C., Zhang, G., et al.: Estimation of forest parameter and biomass for individual pine trees using airborne LiDAR. Trans. Chin. Soc. Agric. Mach. **44**(9), 219–242 (2013)
4. Chen, M., Wan, Y., Wang, M., et al.: Automatic stem detection in terrestrial laser scanning data with distance-adaptive search radius. IEEE Trans. Geosci. Remote Sens. **56**(5), 2968–2979 (2018)
5. Pirotti, F., Guarnieri, A., Vettore, A.: Ground filtering and vegetation mapping using multi-return terrestrial laser scanning. ISPRS J. Photogram. Remote Sens. **76**(2), 56–63 (2013)
6. Calders, K., Newnham, G., Burt, A., et al.: Nondestructive estimates of above-ground biomass using terrestrial laser scanning. Methods Ecol. Evol. **6**(2), 198–208 (2015)
7. Olofsson, K., Holmgren, J., Olsson, H.: Tree stem and height measurements using terrestrial laser scanning and the RANSAC algorithm. Remote Sens. **6**(5), 4323–4344 (2014)
8. Besl, P.J., Mckay, N.D.: A Method for registration of 3-D shapes. In: Robotics - DL Tentative, pp. 239–256. International Society for Optics and Photonics (1992)
9. Chen, Y., Medioni, G.: Object modelling by registration of multiple range images. Image Vis. Comput. **10**(3), 145–155 (1992)
10. Li, J., Wang, Z.M., Ma, Y.R.: Automatic and accurate mosaicking of point clouds from multi-station laser scanning. Geomat. Inf. Sci. Wuhan Univ. **39**(9), 1114–1120 (2014). (in Chinese)
11. Rusu, R.B., Blodow, N., Marton, Z.C., et al.: Aligning point cloud views using persistent feature histograms. In: 2008 IEEE/RSJ International Conference on Intelligent Robots and Systems, pp. 3384–3391. IEEE (2008)
12. Yang, B., Dong, Z., Liang, F., et al.: Automatic registration of large-scale urban scene point clouds based on semantic feature points. ISPRS J. Photogram. Remote Sens. **113**, 43–58 (2016)
13. Zhang, W., Chen, Y., Wang, H., et al.: Efficient registration of terrestrial LiDAR scans using a coarse-to-fine strategy for forestry applications. Agric. For. Meteorol. **225**, 8–23 (2016)
14. Kelbe, D., van Aardt, J., Romanczyk, P., et al.: Marker-free registration of forest terrestrial laser scanner data pairs with embedded confidence metrics. IEEE Trans. Geosci. Remote Sens. **54**(7), 4314–4330 (2016)
15. Henning, J.G., Radtke, P.J.: Multiview range-image registration for forested scenes using explicitly-matched tie points estimated from natural surfaces. ISPRS J. Photogram. Remote Sens. **63**(1), 68–83 (2008)
16. Liang, X., Hyyppä, J.: Automatic stem mapping by merging several terrestrial laser scans at the feature and decision levels. Sensors **13**(2), 1614–1634 (2013)
17. Brenner, C., Dold, C., Ripperda, N.: Coarse orientation of terrestrial laser scans in urban environments. ISPRS J. Photogram. Remote Sens. **63**(1), 4–18 (2008)

Analyzing Spatiotemporal Characteristics of Taxi Drivers' Cognition to Passenger Source Based on Trajectory Data

Zihao Wang[1] 📵, Jun Li[1,2(✉)] 📵, Yan Zhu[1] 📵, Zhenwei Li[1] 📵, and Wenle Lu[1] 📵

[1] College of Geoscience and Surveying Engineering, China University of Mining and Technology, Beijing, Beijing 100083, China
junli@cumtb.edu.cn

[2] State Key Laboratory of Resources and Environmental Information System, Chinese Academy of Sciences, Beijing 100101, China

Abstract. Seeking passengers is a kind of behavior of taxi drivers with clear purposes. They always need to make decisions on where to seek the next passenger after finishing a trip. Experienced drivers are capable to capture passenger source within a short time to reduce no-load time. Most of the existing literatures focus on simulating or analyzing movement patterns of taxi drivers. This research proposes a method of analyzing spatiotemporal characteristics of taxi drivers' cognition to passenger source. Using a seven-day taxi trajectory data set collected in Beijing, an index CLPS is introduced to evaluate taxi drivers' cognitive level to passenger source. Based on this, spatiotemporal distribution of top drivers' cognition to passenger source is explored. The results of the research show that top drivers' cognition to passenger source has obvious spatiotemporal distribution features. This research is expected to provide new ways for understanding human spatial cognition.

Keywords: Spatial cognition · Passenger source · Cognitive level · Taxi driver · Trajectory data

1 Introduction

As an important component of urban public passenger transportation, taxis offer convenient and personalized travel service for urban residents. Urban residents have a wide demand for taxi travelling. Take Beijing, the capital of China, as an example. In 2012, the total operating mileage of taxis was 5.85 billion kilometers, and the passenger volume was 699 million. Therefore, it is of great significance to study the travelling of urban taxis. Due to large scale and long operation time of taxis, they have become one of the main factors causing congestion. In addition, people's demand for convenient travel is becoming more and more urgent. How to make taxis match well with passengers and improve taxis operation efficiency are hot topics in the current stage.

With the widespread use of GPS and wireless communication technology, the trajectory data of urban taxis can be collected in large quantities and the research concerning

S. Di Martino et al. (Eds.): W2GIS 2020, LNCS 12473, pp. 90–94, 2020.
https://doi.org/10.1007/978-3-030-60952-8_9

passenger seeking behavior of taxi drivers are based on such dataset. Existing research mainly includes the following three aspects: (1) give suggestions to taxi drivers by studying their distribution characteristics [1, 2]; (2) find areas with high-quality passenger source by extracting behavior patterns of taxi drivers [3, 4]; (3) simulate passenger seeking behavior of taxi drivers and explore superior methods of passenger seeking by building forecasting models [5, 6].

Most existing research focused on exploring the relationship between taxi drivers and passenger source based on geographic distribution of passenger source, but not investigating how taxi drivers understand the distribution of passenger source from taxi drivers themselves. Some research studied drivers' cognition from the perspective of psychology and neurobiology [7, 8], but ignored their cognition in geographical sense.

Using the trajectory data obtained from GPS devices, this research proposes an index CLPS to reflect taxi drivers' cognitive level to nearby passenger source and classifies top drivers with superior cognitive level. Based on this, the spatiotemporal characteristics of cognitive level of top drivers are explored. The research is expected to provide support for taxi dispatching, intelligent traffic management and location-based service operators.

2 Methods

Firstly, the passenger-seeking segments are separated from the original trajectory data, and the starting points of each seeking segment are extracted. The seeking segments with the cursing time is less than 1 min or longer than 1 h are removed considering the randomness and difficulties to reflect drivers' cognition. Secondly, the study area is discretized into 1 km × 1 km grids for further analysis.

Taxi drivers' cognitive level to nearby passenger source, to a large extent, can be reflected by their passenger-seeking behavior [9]. After a taxi driver finishes an order, he needs to decide where he is going to seek the next passenger based on his comprehensive cognition to the location of passenger source. However, due to different degrees of development in different regions, taxi divers' average seeking time is largely affected by the place they are located at. In order to eliminate the difference of seeking time caused by spatial disparity, the driver's cognition to nearby passenger source is evaluated by comparing a driver's seeking time with the average seeking time of all drivers initiated from the same place.

The cognitive level to passenger source index (CLPS) is calculated as follows:

$$\text{CLPS} = \frac{t_{average}}{t_{ID}}$$

where t_{ID} is a driver's passenger seeking time at a place, and $t_{average}$ is the average seeking time of all drivers initiated from the same place.

The larger the CLPS value of a taxi is, the higher the taxi driver's cognitive level to nearby passenger source is, and vice versa. Meanwhile, the CLPS is used to classify taxi drivers. When the CLPS of a seeking segment is larger than n_1, the seeking behavior is defined as an efficient seeking behavior. When a driver's efficient seeking behaviors account for more than $n_2\%$ of the total behaviors, the driver is defined as a top driver.

3 Results

Beijing is selected as the study area, and the taxi trajectory dataset collected during November 1st to 7th in 2002 are used as study dataset. After screening out noisy data and classifying taxi drivers, there are 65,186 efficient seeking behaviors and 635 top drivers. In addition, the standard deviation of top drivers' CLPS between seven days is 0.1676 in average, while that of non-top drivers is 0.2774, which reflects that the top drivers group has a more significant cognitive mode to passenger source.

3.1 Temporal Distribution Characteristics

In order to explore the change of top drivers' cognitive level to passenger source over time, the CLPS index is calculated by hour. Figure 1 shows the change of CLPS of top drivers during seven days of a week.

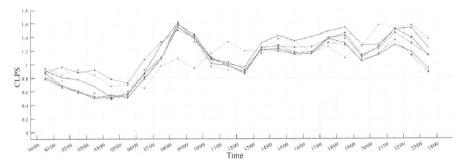

Fig. 1. Top drivers' cognitive level correlated with time period

(1) The top drivers' cognitive level to passenger source in the seven days from November 1st to 7th has a similar pattern. In addition, top drivers have a cognitive mode that changes with time period.

(2) Top drivers have a stronger cognition of the source of passengers in daytime, and a weaker cognition in nighttime. According to the operation mode of the taxi shift system in Beijing [10], it can be inferred that compared with the drivers working in the daytime, the cognitive level of the taxi drivers at night for the same car is generally lower.

(3) In terms of the daytime, top drivers' cognitive level to passenger source shows an increasing tendency from 5:00 to 7:00, and reaches the best cognitive state at around 8:00, 18:00 and 21:00. These are peak hours for people to go to work, leave work and leave work after overtime working, which are also the time when taxis are in a great demand. This reflects that top drivers usually have a sensitive cognition to passenger source than other drivers when the number of passengers is large.

3.2 Spatial Distribution Characteristics

The average CLPS in each grid is calculated based on the CLPS of all seeking behaviors in a grid as shown in Fig. 2, which can help explore the cognitive superiority and weakness regions. From green to red, the color change represents the increment of CLPS of top drivers and top drivers' cognitive level to passenger source (Fig. 2a). Furthermore, based on the passenger seeking trajectory segments, we use the density analysis tool in ArcGIS to explore the seeking area of top drivers (Fig. 2b).

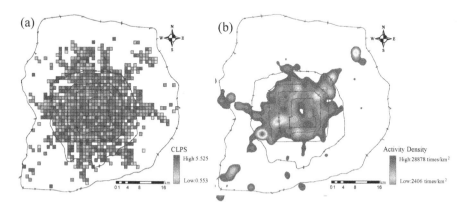

Fig. 2. (a) CLPS in the unit of 1 km × 1 km (b) Top drivers' active regions of passenger seeking behavior (Color figure online)

It can be seen from Fig. 2a that top drivers' cognitive superiority regions account for the vast majority while the weakness regions distribute mainly around the fourth and fifth rings. Compared with top drivers' active regions of passenger seeking behavior, it can be seen that top drivers tend to seek passengers in the regions where they have a strong cognitive level to passenger source. However, it exists an opposite phenomena in the southwest between the fourth and the fifth ring.

4 Conclusion

In this research, an index indicating driver's cognitive level to passenger source is defined. Then, the spatiotemporal characteristics of taxi drivers' cognition to passenger source are explored based on taxi trajectory data.

By analyzing top drivers' cognitive level with time changes, it is found that top drivers' cognition is correlated to peak hours. Furthermore, taxi drivers usually seek passengers in regions where they have a strong cognitive level based on analyzing their spatial characteristics.

Acknowledgement. This research was funded by National Natural Science Foundation of China (No. 41971355), the Open Project of State Key Laboratory of Resources and Environmental Information System and Yueqi Young Scholar Project of China University of Mining and Technology at Beijing.

References

1. Yamamoto, K., Uesugi, K., Watanabe, T.: Adaptive routing of multiple taxis by mutual exchange of pathways. Int. J. Knowl. Eng. Soft Data Paradigms **2**(1), 57–69 (2010). https://doi.org/10.1504/IJKESDP.2010.030466
2. Gong, L., Liu, X., Wu, L., Liu, Y.: Inferring trip purposes and uncovering travel patterns from taxi trajectory data. Cartogr. Geogr. Inf. Sci. **43**(2), 103–114 (2016). https://doi.org/10.1080/15230406.2015.1014424
3. Hu, X., An, S., Wang, J.: Exploring urban taxi drivers' activity distribution based on GPS data. Math. Probl. Eng. 1–13 (2014). https://doi.org/10.1155/2014/708482
4. Zhang, Z., He, X.: Analysis and application of spatial distribution of taxi service in city sub-areas based on taxi GPS data. In: ICCTP 2011: Towards Sustainable Transportation Systems, pp. 1232–1243 (2011). https://doi.org/10.1061/41186(421)121
5. Gao, Y., Jiang, D., Yan, X.: Optimize taxi driving strategies based on reinforcement learning. Int. J. Geogr. Inf. Sci. **32**(8), 1677–1696 (2018). https://doi.org/10.1080/13658816.2018.1458984
6. Zhao, L., Song, Y., Zhang, C., Liu, Y., et al.: T-GCN: a temporal graph convolutional network for traffic prediction. IEEE Trans. Intell. Transp. Syst. 1–11 (2019). https://doi.org/10.1109/tits.2019.2935152
7. Walker, G.H., Stanton, N.A., Young, M.S.: An on-road investigation of vehicle feedback and its role in driver cognition: implications for cognitive ergonomics. Int. J. Cogn. Ergon. **5**(4), 421–444 (2001). https://doi.org/10.1207/S15327566IJCE0504_4
8. Dong, W., Liao, H., Roth, R.E., Wang, S.: Eye tracking to explore the potential of enhanced imagery basemaps in web mapping. Cartogr. J. **51**(4), 313–329 (2014). https://doi.org/10.1179/1743277413Y.0000000071
9. Liu, L., Andris, C., Ratti, C.: Uncovering cabdrivers' behavior patterns from their digital traces. Comput. Environ. Urban Syst. **34**(6), 541–548 (2010). https://doi.org/10.1016/j.compenvurbsys.2010.07.004
10. Tang, L.L., Duan, Q., Kan, Z.H., Li, Q.Q.: Study on identification and space-time distribution analysis of taxi shift behavior. J. Geo-Inf. Sci. 167–175 (2017). https://doi.org/10.3724/sp.j.1047.2017.00167

Comparison of Buildings Extraction Algorithms Based on Small UAV Aerial Images

Haolin Wu, Gaozhong Nie[(✉)], and Xiwei Fan

Institute of Geology, China Earthquake Administration, Beijing 100029, China
Niegz@ies.ac.cn

Abstract. Using small rotor UAV, this paper acquired the images of qionghalajun village of the artushi city in Xinjiang autonomous region. Then, two image analysis methods of object-oriented and pixel-oriented methods, respectively, were used to extract the building of the study area. It results show that the object-oriented method can effectively remove the impact of salt and pepper noise on classification and ensure the integrity of the shapes of the buildings. But the impact of different objects with similar spectral and texture information will decrease the classification accuracy. In the pixel-oriented extraction method, the improved mathematical morphology algorithm is added, which can effectively improve salt and pepper noise, maintain the continuity and integrity of the edge of the building, and solve the problem of partial farmland and houses in the object-oriented method.

Keywords: UAV · Buildings extraction · Pixel-oriented · Object-oriented

1 Introduction

The pre assessment of earthquake disaster loss is one of the key points to carry out earthquake emergency command timely and scientifically. 80% of the casualties caused by the earthquake are caused by the destruction of houses. In the process of earthquake disaster loss pre assessment, the key is to do a good job in the survey of housing construction status and population scale in the assessment area [3, 4]. And for the study area of housing construction extraction work is a strong guarantee of the above survey.

Because unmanned aerial vehicle (UAV) remote sensing technology has many advantages compared with satellite remote sensing technology [1, 2], in the actual work of earthquake disaster loss pre assessment, UAV gradually began to replace the traditional satellite remote sensing technology to obtain the information of surface buildings. In the process of investigation, because there are many investigation points in each area, and the investigation time of each investigation point is limited, most of the UAVs carried are small rotor UAVs with the advantages of short flight preparation time, flexible operation, easy to carry and so on. The sensors of the UAV are mostly digital cameras. Although the image resolution is high, the spectral resolution is very low, that is to say, the number of bands of the returned image data is small. In this paper, two image analysis methods, object-oriented and pixel oriented, are used to extract the buildings in the study area, and the results of the two algorithms are compared, and the advantages of the two methods are analyzed under the data condition

© Springer Nature Switzerland AG 2020
S. Di Martino et al. (Eds.): W2GIS 2020, LNCS 12473, pp. 95–101, 2020.
https://doi.org/10.1007/978-3-030-60952-8_10

2 Study Area and Data Acquisition

The research area selected in this paper is qionghalajun village, Halajun Township, artushi City, Xinjiang. In this paper, a small four rotor UAV is used, which is a professional version of Dajiang spirit 3. The effective pixels of the camera are 12.4 million. In qionghalajun village, Xinjiang, the image data are acquired automatically by control software with the side direction overlap of 75% and the course overlap of 50%. 69 high-resolution aerial images are obtained with the flight height (H) is set to 100 m ground, because the UAV camera focal length (H = 4 mm) and sensor size (d = 0.001 58 mm) are known. Reference formula:

$$R = \frac{H * d}{h} \tag{1}$$

in which: R is the ground spatial resolution. The ground spatial resolution of UAV image is about 3.9 cm, which is better than that of space platform visible light remote sensing image, and reaches the standard of high-resolution remote sensing image. After preprocessing, registering and stitching 69 UAV images, the (Digital Orthophoto Map) DOM image of the research area is finally generated (as shown in Fig. 1).

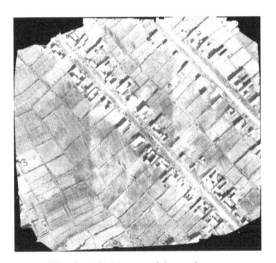

Fig. 1. DOM image of the study area

3 Experimental Process

3.1 Object Oriented Image Analysis Method

Firstly, we get the heterogeneity F of each image object in each segmentation, and use the region merging algorithm based on the principle of minimum heterogeneity (that is, the idea of multi-scale segmentation) to segment the image. Firstly, a single pixel is merged into a small object, and then the small object can be merged into a large object gradually through several steps. The process of merging must ensure that the heterogeneity of image object is less than the given threshold.

In the actual segmentation process, due to the small difference in the spectral characteristics of the house building, the weight of the spectral factor is set to 0.2, the weight of the shape factor is 0.8, because the visual observation shows that the Roof Texture of the house is relatively smooth, the setting of the smoothness weight is 0.7, the tightness weight is 0.3, and The scale parameter of heterogeneity is set to 350 after many attempts is the best. Start the segmentation with any pixel as the center, and calculate whether the heterogeneity F of adjacent pixels is less than the given threshold s by formula (2) after the first segmentation. If the threshold conditions are not met, carry out the second segmentation based on the last segmented image object, and repeat the above steps until the heterogeneity of the two adjacent image objects is less than the given threshold s to complete the final segmentation. The segmentation results using multi-scale segmentation are shown in Fig. 2a.

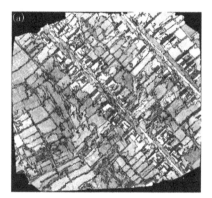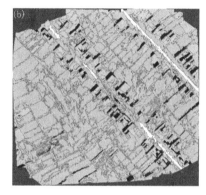

Fig. 2. Image segmentation results (a) and object-oriented classification results (b) (Color figure online)

For the segmented image, the gray mean and standard deviation of three bands of each image object are extracted as spectral features. The angular second moment (ASM) and inverse different moment (IDM) values are calculated by using the gray level co-occurrence matrix inside each object, and the 1:1 weighted sum is taken as texture features. ASM value reflects the uniformity of gray distribution and texture thickness in each object, and the calculation formula is as follows:

$$ASM = \sum_{i=1}^{k} \sum_{j=1}^{k} [G(i,j)]^2 \qquad (2)$$

IDM reflects the homogeneity of image texture and measures the local changes of image texture. The calculation formula is as follows:

$$IDM = \sum_{i=1}^{k} \sum_{j=1}^{k} \frac{G(i,j)}{1 + (i-j)^2} \qquad (3)$$

The extracted feature vectors are used to classify the segmentation results, as shown in Fig. 2b, where green represents farmland, blue represents buildings, yellow represents roads, and black represents shadows.

3.2 The Realization of Pixel Oriented Morphological Processing Algorithm

The frequency histogram of gray-scale image is used to carry out the piecewise linear pull-up of gray-scale value. The image with obvious wave crest (between 250 and 255) of gray-scale value is extracted as the foreground image. The method of mathematical morphology image processing is used to remove the complex background in the image. The specific operations are as follows: 1. considering that the unrelated pixels are all fine spots, the area of a few consecutive unrelated pixels is also far small In the fact of building area, the long horizontal line is used as the opening operation of structural element B for reconstruction. In the reconstruction process, the image is first corroded, just like the standard morphological opening operation. However, the corroded image is used as the marker image f in the image reconstruction process, and the original image is used as the template g to perform the morphological reconstruction process again: Initialize tag image f to h1(Tag f must be a subset of G). Building structural elements B = ones(1,11), To define connectivity, we use 8-connectivity. Repeat $h_{k+1} = (h_k) \oplus B \cap G$ until $h_{k+1} = h_k$, Completion of reconstruction operation. 2.Perform the top hat reconstruction on the reconstructed image, that is, subtract the reconstruction operation from the original image, and the execution result is recorded as f-thr. 3.The top cap of the original image is reconstructed by changing the structural element into a long vertical line B = ones(11, 1), and the reconstruction result is recorded as f-thi. 4.f-thr is used as the template, min(f-thi, f-thr) is used as the marker, and morphological reconstruction is performed as the final denoising result. Canny operator is used to detect the edge of the reconstructed image, and the threshold vector T is set as 0.04 to 0.1, The standard deviation sigma of the smoothing filter is set to 0.1.

In this paper, based on the expansion method of reconstruction, the iterative method is used to open the operation, accurately restore the shape of the corroded object. In practice, because most of the houses in the study area are rectangular, and the area of the houses is greater than 5 m^2, the rectangular structural element is selected and the threshold is set to 5 when the operation is carried out. The image is opened by iterative morphological operation, the holes and cracks in the house are filled, and the internal shape of the house is reconstructed. Then, the image is closed by iterative morphological operation using the size of 5em rectangular structural elements, and the external contour of the house is reconstructed to remove the fine protrusion and make the house boundary flat. Finally, the reconstructed image is filled. The edge detection results after morphological processing are shown in Fig. 3, and the final extraction results are shown in Fig. 4.

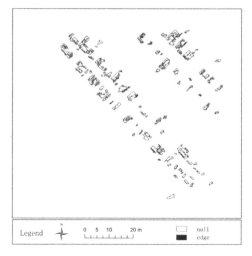

Fig. 3. Edge detection results after adding morphological denoising

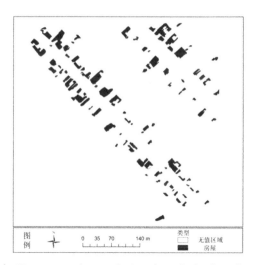

Fig. 4. House extraction results based on pixel-oriented method

3.3 Accuracy Evaluation

The confusion matrix is used to evaluate the classification results: the results of object-oriented house extraction and mathematical morphology are used as the classification data respectively, and the house location (Fig. 5) obtained by vectorization of the research area with DOM as the base map is used as the real surface data to establish the confusion matrix, and the mapping accuracy, user accuracy and missing error are calculated And misclassification error.

The calculation results are shown in Table 1.

Fig. 5. Real data on surface

Table 1. Accuracy evaluation of classification results

Classification method	Cartographic accuracy	User accuracy	Misclassification error	Leakage error
Object oriented house extraction	73.56%	58.43%	41.57%	26.44%
Pixel oriented morphological house extraction	81.42%	76.54%	23.46%	18.58%

4 Conclusions

The object-oriented method can effectively remove the impact of salt and pepper noise on classification; in the process of segmentation, adding the shape and texture information of ground objects reasonably can keep the edge of each object continuous after segmentation, which also ensures the integrity of the house shape during image classification. However, some areas are misclassified in the results. It can be found that most of the misclassified areas are farmland. This area not only has similar spectral characteristics with the house, but also has similar texture and rectangular shape with the house roof. Because the object-oriented classification is based on the classification of each image object, while in the multi-scale image segmentation, only the heterogeneity difference between adjacent image objects can be considered, and the phenomenon of feature similarity may occur if the non adjacent objects cannot be considered. Because there are many features similar between farmland and houses in this area, if the object-oriented method is used in this area In this way, the internal features of image objects can be extracted and classified, which will easily lead to the misclassification of image objects.

However, using pixel oriented method to separate houses will lead to the problem that the integrity of the ground objects extracted is not high and salt and pepper noise is easy to appear. In the process of experiment, first of all, we need to consider how to eliminate the noise without affecting the real house pixels, and in the process of house extraction, we need to keep the shape integrity of houses as much as possible. In this paper, the mathematical morphology method proposed by the predecessors is improved, and the pixel level house extraction is better realized, which solves the problem that some farmland and houses are misclassified due to the similarity of internal characteristics of image objects in the object-oriented method.

Acknowledgements. This work was jointly supported by the National Key R&D Program of China (No. 2018YFC1504403 and No. 2018YFC1504503), the National Natural Science Foundation of China (Grant No. 41601390), and the China Earthquake Administration Special Project Surplus Fund (High Resolution Rapid Post-Earthquake Assessment Techniques)

References

1. Berra, E.F., Gaulton, R., Barr, S.: Use of a digital camera onboard a UAV to monitor spring phenology at individual tree level. In: Geoscience and Remote Sensing Symposium (IGARSS), pp. 3496–3499 (2016)
2. Dandois, J.P., Ellis, E.C.: High spatial resolution three-dimensional mapping of vegetation spectral dynamics using computer vision. Remote Sens. Environ. **136**, 259–276 (2013)
3. Geiß, C., Pelizari, P.A., Marconcini, M.: Estimation of seismic building structural types using multisensor remote sensing and machine learning techniques. ISPRS J. Photogramm. Remote Sens. **104**, 175–188 (2015)
4. Feng, T., Hong, Z., Wu, H.: Estimation of earthquake casualties using highresolution remote sensing: a case study of Dujiangyan city in the May 2008 Wenchuan earthquake. Nat. Hazards **69**(3), 1577–1595 (2013). https://doi.org/10.1007/s11069-013-0764-1

Mapping the Earthquake Landslide Risk, A Case Study in the Sichuan-Yunnan Region, China

Xu Jinghai[⊠], Bu Lan, Li Bo, and Zhou Haijun

College of Geomatics Science and Technology, Nanjing Tech University, Nanjing 211816, China
Xu_jing_hai@163.com

Abstract. Rapid and effective assessment of the landslide risk and its spatial distribution after the earthquake provide good support for earthquake emergency rescue. In this paper, we develop a landslide risk map in the Sichuan-Yunnan region for post-earthquake emergency rescue with grid technology based on historical seismic landslide data of Wenchuan earthquake. The historical data of Wenchuan earthquake is extracted based on the Aster images. The seismic landslide risk assessment model is introduced based on logistic regression model, in which seismic intensity and slop of digital elevation model are selected as influence factors (input parameters). Then we grid the landslide risk estimation model to generate landslide risk maps in resolution of 90 m. So the risk maps are prepared before earthquake, the landslide risk can be quickly estimated after earthquake according to few information acquired from earthquake monitor network, such as magnitude, focal depth, earthquake location. Finally, a seismic landslide risk assessment system was developed to realize the application of the risk maps.

Keywords: Earthquake landslide · Risk assessment · Logistic regression · Gridding technology

1 Introduction

Statistics on earthquakes with magnitude MS ≥ 5.0 since the 1949 show that earthquake induced landslides are occurred located in most part of provinces in China, especially in western mountainous regions [1]. Research on the danger of seismic landslides has been a subject, which is worth of attention in the field of seismic engineering and has been extensively studied [2–8]. Parts of the researches are based on the map of seismic intensity zoning in China. Based on that, the landslide hazard distribution map is developed by introducing a landslide hazard assessment model. For example, the literature [9, 10] has established network diagram of earthquake landslide danger zones in Yunnan Province and the country. Research in this area can provide earthquake risk distribution before the earthquake. Thus it provides reference for disaster mitigation. In order to provide a good reference for disaster prevention in the region in the future, some researchers have drawn landslide danger maps based on landslides in historical earthquake cases and specific earthquake cases [11]. In general, existing studies focus more on the hazard

© Springer Nature Switzerland AG 2020
S. Di Martino et al. (Eds.): W2GIS 2020, LNCS 12473, pp. 102–107, 2020.
https://doi.org/10.1007/978-3-030-60952-8_11

distribution of landslides in the future earthquake (pre-earthquake) and can provide good support for disaster mitigation preparation. Therefore, the spatial distribution of landslide hazard maps usually generated by such studies is not high, such as in cities or regions. In fact, the landslide risk analysis after the earthquake has important significance and requirements for disaster reduction. For example, if the landslide risk and its relatively fine spatial distribution (such as 90 m) can be quickly evaluated in the initial stage after the earthquake, it can provide a reference for disaster relief departments to reduce the secondary earthquake damage caused by following landslide. In this paper, taking the Wenchuan earthquake as background, grid technology is introduced to develop a landslide hazard distribution map for post-earthquake emergency rescue by spatializing the traditional landslide risk assessment model.

2   Seismic Landslide Risk Assessment Model

At present, the models of evaluating seismic landslide risk including its spatial distribution are widely studied. We classified the models into two types: one is the model based on physics and engineering geology. It is based on the physical mechanism of landslide occurrence and uses mechanical models to quantitatively express the stability status of slopes, such as the model of limit equilibrium of slopes, Newark displacement analysis, and numerical simulation [12–14]. The establishment of this kind of model requires the collection of various data on geology, geomorphology, river hydrology, etc. The kind of model is more suitable for the monitoring and evaluation of single landslide. Another kind of model is based on the analysis of historical landslide data. The landslide risk in future can be deduced by analyzing the influence factors of historical landslide instability, and relying on empirical and mathematical algorithms, such as historical regression method, expert scoring method, etc. [15, 16]. The model considers the occurrence principles of landslide as a black box, so this kind of model is more suitable for regional landslide risk analysis and assessment. In this paper, the second kind model will be used to evaluate the risk of seismic landslides in the Sichuan-Yunnan region. The process of building model is shown in the Fig. 1. The model aims at the quickly response after earthquake. With the period, the information about earthquake is reliable.

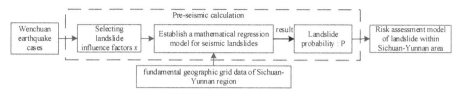

Fig. 1. The process of building risk assessment model of seismic landslide in Sichuan-Yunnan region

Considering the key purpose of post-earthquake emergency rescue, we select the slope and seismic intensity as the influencing factors of earthquake landslides. A logistic regression model is used to establish the model in the Sichuan-Yunnan region. The formula is

$$P = \frac{1}{1 + \exp(-\alpha - \sum_{n=1}^{n} \beta_n x_n)} \tag{1}$$

Where P is the probability of occurrence of landslide, x_n are the influence factors of seismic landslide; β_n are the regression parameters; α is a constant term.

3 Landslide Risk Map Development

3.1 Extracting Disaster of Wenchuan Earthquake

Wenchuan is located in Sichuan province, about 130 km from Chengdun– the capital of Sichuan Province. There are the Minjiang River passing by and an active area of the Longmen fault zone. In May 12th, 2008, a giant earthquake (Ms. 8.0) is occurred, which covers almost half China and Sichuan and Yunnan provinces are the most serious disaster area. The earthquake causes 69227 people death and thousands seismic landslides. We collect seismic landslide data from NASA's ASTER sensor, which has a wider spectral range with high-resolution images in 15 m. At present, users can apply for three levels of data products L1, L2, and L3. We download the ASTER_L3T images within two years after May 2008, that cover the disaster area of Wenchuan earthquake.

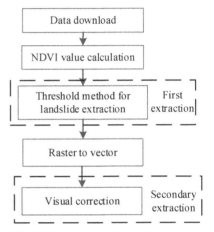

Fig. 2. The process of extracting landslide

The extraction of seismic landslide data is based on three visible light bands of the ASTER image. Due to the relatively small number of bands, the test find that just using supervised or unsupervised classification methods could not achieve good results. So we use the normalized difference vegetation index to make preliminary recognizing and then extract the landslides by visual interpretation as shown in Fig. 2. The normalized vegetation index is a comprehensive reflection of the type of vegetation, coverage form, and growth conditions in a unit pixel. The expression of the normalized difference vegetation index (NDVI) is:

$$NDVI = \frac{NIR - R}{NIR + R} \tag{2}$$

Among them, NIR stands for the near-infrared channel (0.7 to 1.1 μm) in the multi-spectral remote sensing image, and R stands for the visible red band (0.6 to 0.7 μm). Because it is sensitive to green vegetation, it is often used to detect the growth status of vegetation, vegetation coverage and elimination of some radiation errors. A negative value of NDVI indicates that the ground is covered cloud, water, snow, etc.; the value 0 indicates that there are rocks or bare soil; positive values indicate that there is vegetation coverage.

After the NDVI value is calculated, the preliminary landslide extraction is performed by ArcMap software based on the threshold range. The empirical threshold is generally in the range of 0–0.12. It is worth noting that the impact of earthquakes on vegetation is different due to the different seismic intensity zones. Thus the threshold range needs to be modified according to seismic intensity. After the preliminary extraction, manual visual interpretation is required to exclude the area that affected by rivers, dammed lakes, and farmland. Finally, more than 100,000 landslides are extracted and we eliminate the landslide area less than 1 km^2. The seismic landslides of Wenchuan earthquake are shown in Fig. 3.

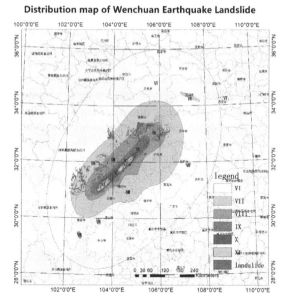

Fig. 3. Distribution map of Wenchuan earthquake landslide

3.2 Gridding Landslide Risk Assessment Model

Data of the model include the Wenchuan earthquake intensity distribution map, the slope map from SRTM90 m (http://srtm.csi.cgiar.org/), and the historical landslide vector map. We grid the data and corresponding model to develop grid landslide risk data for Sichuan-Yunnan region. The main steps are shown in Fig. 4.

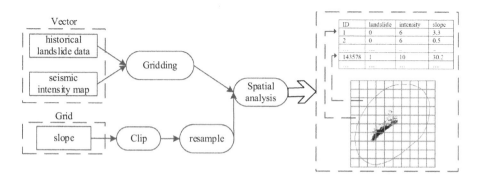

Fig. 4. Main steps of gridding the model

4   Application of the Landslide Risk Data

We develop seismic landslide risk evaluation system with the support of the landslide risk data. The system has the function to analyze the seismic influence range, estimate seismic landslide risk for emergency response. As shown in Fig. 5, risk dasymetric map of seismic landslide in Sichuan-Yunnan region (take the intensity VI as an example) has been calculated. After earthquake the system can quickly estimate the possible landslide risk by generating seismic intensity map according to few information (earthquake magnitude, focal depth, earthquake location), which almost the only information just after earthquake (i.e. 1 h after earthquake). The landslide map will provide good guidance to emergency response and rescue.

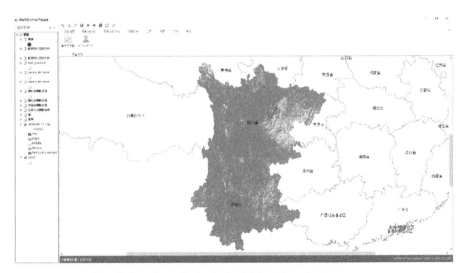

Fig. 5. Interface of seismic landslide risk evaluation system

5   Conclusion

How to quickly estimate possible landslide risk caused by earthquake within short time after earthquake is a very useful and challenging work for earthquake emergency response and rescue. We aim to quickly estimate seismic landslide risk for emergency response and rescue in this study. Logistic regression model is selected, in which seismic intensity and slope are used as input parameters. The historical landslide data of Wenchuan are used to make regressive analysis. Then seismic landslide maps and corresponding evaluation system are developed with gridding method. So the evaluation process of seismic landslide is simplified and improve the effectiveness of the evaluation. We will further optimize the evaluation model and process in the future.

Acknowledgements. This work was jointly supported by The National Key Research and Development Program of China (Grant No. 2018YFC0704305, No. 2018YFC1504503), and the Science and Technology Project of Jiangsu Construction System in 2018 (Grant No. 2018JH011).

References

1. Xiuying, W.: Study of fast evaluation technique of earthquake-induced landslides and their effects on earthquake emergency rescue. Recent Dev. World Seismolog. **29**(7), 2159 (2011)
2. Akgun, A.: Application of remote sensing data and GIS for landslide risk assessment as an environmental threat to Izmir city (west Turkey). Environ. Monit. Assess. **184**(9), 5453–5470 (2012). https://doi.org/10.1007/s10661-011-2352-8
3. Jibson, R.W.: A method for producing digital probabilistic seismic landslide hazard maps. Eng. Geol. **58**(3), 271–289 (2000)
4. Romeo, R.: Seismically induced landslide displacements: a predictive model. Eng. Geol. **58**(3), 337–351 (2000)
5. Edwin, L.: Landslide inventories: the essential part of seismic landslide hazard analyses. Eng. Geol. **122**(1), 9–21 (2010)
6. Jibson, R.W.: Regression models for estimating coseismic landslide displacement. Eng. Geol. **91**(2), 209–218 (2007)
7. Xu, C., Xu, X.: The spatial distribution pattern of landslides triggered by the 20 April 2013 Lushan earthquake of China and its implication to identification of the seismogenic fault. Chin. Sci. Bull. **59**(13), 1416–1424 (2014). https://doi.org/10.1007/s11434-014-0202-0
8. Tian, Y.: Inventory and spatial distribution of landslides triggered by the 8th august 2017 M_W6.5 Jiuzhaigou earthquake, China. J. Earth Sci. **30**(1), 206–217 (2018). https://doi.org/10.1007/s12583-018-0869-2
9. Tang, C.: Study on prediction of earthquake induced landslide hazard area supported by GIS. J. Earth Seismol. Res. **24**(1), 73–81 (2001)
10. Zhu, L.: Risk assessment of landslide in China using GIS technique. Rock Soil Mech. **24**(S2), 221–224 (2003)
11. Jianping, Q.: Study on Wenchuan earthquake-induced landslide risk zonation. J. Eng. Geol. **23**(2), 187–193 (2015)
12. Zhang, J.: Susceptibility of landslide based on artificial neural networks and fuzzy evaluating model. Sci. Surveying Mapp. **37**(3), 63–66 (2012)
13. Wang, Y.: Occurrence probability assessment of earthquake-triggered landslides with Newmark displacement values and logistic regression: The Wenchuan earthquake. China. Geomorphol. **258**(258), 108–119 (2016)
14. Zhang, J.: The numerical and sensitivity analysis of the landslide induced by earthquake. Sci. Technol. Eng. **21**(1), 16–24 (2013)
15. Zhao, J.H.: Comparison of models for hazard assessment of landslide. J. Nat. Disaster **15**(1), 128–134 (2006)
16. Zhen, Y.: Geological hazard risk assessment in Hongkou county of Dujiangyan. Sci. Surveying Mapp. **41**(11), 66–71 (2016)

Crowdsourcing, Volunteered Geographic Information and Social Networks

A Focused Crawler for Web Feature Service and Web Map Service Discovering

Víctor Macêdo Alexandrino[1]([✉]), Giovanni Comarela[1,2],
Altigran Soares da Silva[3], and Jugurta Lisboa-Filho[1]

[1] Departamento de Informática, Universidade Federal de Viçosa,
Viçosa, MG 36570-900, Brazil
{victor.alexandrino,jugurta}@ufv.br, gc@inf.ufes.br
[2] Departamento de Informática, Universidade Federal do Espírito Santo,
Vitória, ES 29075-910, Brazil
[3] Instituto de Computação, Universidade Federal do Amazonas,
Manaus, AM 69080-900, Brazil
alti@icomp.ufam.edu.br

Abstract. Due to globalization and the technological advances of the last decades, a large amount of data is created every day, especially on the Web. Web Services are one of the main artifacts created on the Web; they can provide access to sources of data of many sizes and types. In this work, we approach the challenge of designing and evaluating a Web Crawler to find two OGC (Open Geospatial Consortium) web services standards: WFS (Web Feature Service) and WMS (Web Map Service). Commercial search engines, e.g., Google and Bing, are indubitably doing a useful job as general-purpose search engines. However, some applications require domain-specific search engines and Web crawlers to find detailed information on the Web. Therefore, this work presents *Spati-Harvest*, a Web crawler that focuses on the task of discovering WFSes and WMSes. SpatiHarvest is a combination of some of the most advanced techniques found in the literature. Through experiments, we demonstrate that it is capable of finding more WFSes and WMSes, with less effort, when compared to the state of the art.

Keywords: Geographic information retrieval · Web crawling · Web services · Open Geospatial Consortium

1 Introduction

Spatial information has been widely used in different types of systems and applications, mainly due to the increasing facility of obtaining this kind of data. Spatial information is useful to answer questions and support the decision making in different areas, such as: environment, climate, urban traffic, geo-marketing, a fast reaction in natural disaster cases, among others. Many private and governmental institutions get large amounts of data of such nature, which little by little are made available through products and services, such as web services [1].

© Springer Nature Switzerland AG 2020
S. Di Martino et al. (Eds.): W2GIS 2020, LNCS 12473, pp. 111–124, 2020.
https://doi.org/10.1007/978-3-030-60952-8_12

Studies show that a large volume of non-indexed pages, even by large search engines, is present on the Web [2]. Therefore, due to the variety of content and the high volume of information on the Web, it is worth studying and developing Web crawlers (also known as spiders or robots), especially those that are domain-specific. In this context, it is possible to assist the robot to explore the domain more extensively by applying techniques such as BFC (Bootstrap Focused Crawler) [3], filters, and supervised learning. This paper considers the domain of Web Services under the WFS (Web Feature Service) and WMS (Web Map Service) standards, which are both for providing spatial data interoperability and maintained by the OGC (Open Geospatial Consortium). More specifically, our goal is to build a Web crawler focused on discovering Web pages that can be or contain URLs (Uniform Resource Locators) related to WFSes or WMSes. In this paper, these Web services are also referred to as *spatial information sources*.

Due to search engines, the problem of discovering WFSes and WMSes is partially solved, once the surface Web is constantly scanned and indexed. However, the surface layer of the Web is estimated to be considerably smaller than its hidden layer [4]. Such a difference is mainly due to two reasons: sometimes, a crawler is not capable of reaching certain pages, for instance, due to the pruning applied in favor of the search space reduction; and many pages are dynamically generated via queries and transactions.

Given that the non-indexed layer of the Web is considerably larger than its surface and that until 2020 the volume of information on the network may reach 44 zettabytes [5], it is a reasonable hypothesis that there may be a significant amount of spatial data sources yet to discover and to index. Search engines, such as, Bing, Google, and Yahoo!, are not specifically designed to find URLs of OGC Services when using basic queries [26]. Furthermore, even resources that are already known may not be so easily located via queries made to those search engines. This is because such resources may be amid a large volume of related, but not relevant search results, once commercial search engines are designed to be general-purpose.

Motivated by the such problematic, this paper presents a Web crawler called *SpatiHarvest*, which was designed by combining techniques and concepts from the Web crawling literature such as bootstrap focused crawling, parallel crawling, and focused Web crawling. *SpatiHarvest*'s goal is to improve the discovery of WMSes and WFSes. We conducted a series of experiments to compare *SpatiHarvest* to the state of the art *PolarHub* [6]. Our results show that *SpatiHarvest* is more efficient and that it is capable of finding more WMSes and WFSes for the same query.

The remaining of this paper is organized as follows: Sect. 2 presents the theoretical foundation behind *SpatiHarvest*. Section 3 presents related work. In Sect. 4, design and implementation are described. In Sect. 5, we present a comparison between *SpatiHarvest* and *PolarHub*. Finally, Sect. 6 presents conclusions, open questions, and directions to future work.

2   Background

In this section, we present the basic concepts, which are necessary to understand *SpatiHarvest*'s architecture. More specifically, we discuss general web crawling, focused web crawling, parallel web crawling, and bootstrap focused crawlers.

2.1   Web Crawler

Web Crawler is a software, also known as robot or spider, built with the purpose of indexing Web pages. The first crawlers appeared almost at the same time as the Internet itself [24]. Such crawlers can be constructed using algorithms such as Depth-First Search and Breadth-First Search. In any case, the search starts from a set of URLs (seeds). Such a set can be chosen arbitrarily or using some method of seed selection. Once a page is discovered and retrieved, its source is parsed and analyzed. At this stage, any URL present in the page's source is also processed in a similar fashion [7].

There are several challenges for implementing a crawler. One of these challenges is choosing which URLs to use during the search. More specifically, given the URLs (links) on a page, which ones should the crawler also explore? In this context, the set of URLs that will be explored and the order of such exploration are relevant. The decision of which URLs should be used in the search can be guided by different methods. Some examples of criteria are: current depth of the search, page's language, domain's country, and, in some cases, the page's topic (or subject). In the latter, supervised learning (e.g. SVM - Support Vector Machines) techniques can be used in order to classify if the URLs contained in a certain page will be added to the frontier of exploration or not [8].

2.2   Focused Web Crawler

A Focused Web Crawler is a regular Web crawler, which aims at indexing only Web pages related to a specific topic of interest. For instance, one can create a Web crawler that focuses on health care pages or academic papers available on the Web. The idea of using crawlers on certain topics is not new. Since 1999, there have been many proposals on reducing the domain of pages to be indexed by a robot. In this context, the robot may not be able to answer any question queried by a user (such as queries outside the scope of interest). However, queries made within the topic explored by the robot can be answered with more up-to-date and relevant information. Moreover, the focused version can considerably save computational resources (e.g., network and disk) since it does not have to explore nor index the whole Web [9].

As for regular Web crawlers, Focused Web Crawling algorithms can be the result of several combinations of techniques, such as Breadth-First Search and Best-First Search [22]. However, there are specific challenges that need to be addressed to achieve efficiency. For instance, a general crawler aims at indexing the whole Web, while a focused crawler has to remain within the topic of interest, otherwise, the results of the query might not be meaningful and computational resources may be unnecessarily used. In addition, the work of a focused crawler starts even before any query is made. Initially, a set of seed pages [10] has to be formed. These pages are necessary to inform the robot where to start exploring. Ideally, the seeds have to enable the discovery of any page on the Web related to the topic of interest. Hence, given a topic, building the right set of seeds is an important challenge by itself.

In particular, a crawler that focuses on WFS and WMS tracking significantly differs from a general one. For instance, the focused crawler should prioritize WFSes and WMSes over regular Web pages, which not necessarily is true in the general case. Furthermore, queries can be performed differently in these two scenarios, i.e., generally, queries are textual, composed by keywords, but in the focused case, they may be based on a set of geospatial attributes.

2.3   Parallel Web Crawler

From an implementation perspective, a crawler may have multiple instances running in parallel, and there are different ways, with different challenges, to do so. Two of the possible architectures for a robot are the distributed and intra-site. In the first, the robot instances are not in the same computer network, i.e., they can be far away, at different facilities, communicating over the Internet. In the second, the robots are in the same local network, i.e., they are usually closer and might communicate over the Intranet [23]. There is also the possibility of running different robot's instances in the same computer, but in different processors, cores, or threads, or even using a Graphic Processing Unit (GPU).

When using a parallel crawler, it is necessary to define some architectural details besides those mentioned above. The robots can function in an independent, static, or dynamic fashion. In the independent mode, each robot's instance works from the seed set itself, without consulting other instances for any purpose, i.e, it is a simpler architecture, but it may suffer from serious problems, such as duplicate page indexing. In the dynamic operation, there is a central controller regulating the work of all other robots, constantly distributing link partitions to each instance of the Crawler. In the static operation, initially each instance of the robot receives a seed partition and then only operates from that partition. Hence, a static architecture does not demand a central controller [23].

2.4   Bootstrapping Focused Crawling

Bootstrapping Focused Crawling (BFC), a framework created with the purpose of supporting focused Web crawlers, is a method to improve the process of automatically finding a good set of seeds, given a query, hence, increasing the reach and harvest rate within a topic [3].

The motivation for using BFC is presented in Fig. 1, where an important challenge that needs to be overcome can be seen. In the figure, the nodes (i.e., pages) of interest (solid blue) are in different connected components. Those components are isolated from each other by non-relevant nodes (dashed red). A crawler that places all the seeds in a single component (e.g., Fig. 1(left)) will not be able to discover all relevant pages via standard methods, such as Breadth-First Search or Depth-First Search. The idea of using BFC is to automatically spread the seeds across several connected components of relevant pages (e.g., Fig. 1(right)).

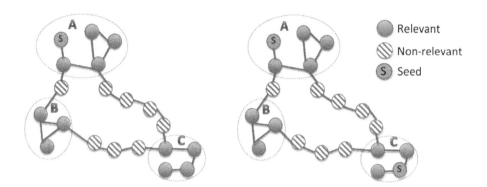

Fig. 1. Graph components connected by irrelevant vertices on the left, BFC selecting seeds among the vertices of two of components on the right [3] (Color figure online)

In order to spread seeds in different components, BFC works as follows: given an input query, (e.g., "Web Feature Services from Australia"), N new queries are built using different keywords related to the same subject of the input. Those keywords are extracted from pages already retrieved on the first query. Then, for each new query, searches on general-purpose search engines are performed in order to obtain a more varied, but still within the scope, set of results. Finally, by joining all the retrieved results, the algorithm has as output a set of seeds that more likely will spread across the Web.

3   Related Work

There is a significant amount of literature related to general-purpose Web crawlers and focused Web crawlers. Specifically in the context of this work, a geospatial focused Web crawler, significantly fewer initiatives were found. Some of those are worth taking note of.

GeoWeb Crawler [11] is a search engine designed with the primary purpose of "providing access to geospatial data or services through the worldwide coverage of the Internet". The algorithm presented in [11] has some characteristics that stand out, such as the use of the map-reduce paradigm, where a master performs the mapping of a set of URLs to workers. Then each worker plays the role of retrieving the page from its URLs. In the end, when each worker has finished its task, a reduce step is performed and the recovered pages are returned to the master. Another important feature is the method used to detect duplicate pages. A bloom filter [12], which allows storing a URL via hash, thus saving memory and processing time. However, such a method does not solve the problem of

Duplicate URLs with Similar Text (DUST) [13], since the duplication analysis is solely based on the comparison of the hashes corresponding to the URLs.

Polarhub is a search engine built to, "discover geospatial data distributed on the Web in an efficient and effective way" [6]. The algorithm developed has, among others, the following characteristics: it uses seeds retrieved from Google, which is a common practice in the literature [14–16], considering that Google Search is able to return relevant generic results from a given input query, thus providing subsidy for a good start of search and indexing; the crawler uses parallel processing to "speed up" the page retrieval process; and it performs a URL rebuild process, in an attempt to create a URL that may correspond to an OGC Web service standard.

A disadvantage of both algorithms mentioned above is that they do not perform any type of pruning when exploring the Web graph. Thus, even though a retrieved page is completely irrelevant (e.g., out of topic), the crawler will still use the URLs contained in this page to continue the search process, as long as the maximum depth allowed is not reached. Hence, processing time and storage may be wasted by analyzing irrelevant pages. Another problem is that no kind of refined technique is applied to the process of building an appropriate set of seeds, i.e., the presented algorithms only retrieve URLs via Google queries and by using the results from those queries, they perform all the search and indexing. As described in the remainder of this manuscript, *SpatiHarvest* is designed to overcome these two main issues.

4   The *SpatiHarvest* Focused Web Crawler

In this section, we present the *SpatiHarvest* architecture and the principles that guided its design. We also discuss the technologies that were used in our proof of concept. We emphasize that, despite our implementation decisions, there are many other viable choices.

4.1   System Architecture

SpatiHarvest's architecture is illustrated in Fig. 2. The system has three main components. The first one is the *BFC* (Bootstrap Focused Crawler), which is responsible for the seed selection. The second, *Crawler Controller*, has the role of a central controller, being responsible for distributing URLs to the workers. The third component, *Crawler Controller Thread*, may have many instances, each one with the same role as a worker node, where most of the crawler's work is effectively performed. More specifically, given a URL sent by the *Crawler Controller*, each *Crawler Controller Thread* is responsible for retrieving the page, parsing the page's source, checking the page's type (i.e., whether the page is a Web Feature Service, Web Map Service, or none of them), and persisting the data into a database.

Figure 2 also reveals the *SpatiHarvest*'s parallel nature. There is a central controller that is in charge of managing and delegating URLs to several workers, which run in parallel. *SpatiHarvest*'s multi-thread architecture can be scaled vertically or horizontally [25], depending on specific implementation decisions and application requirements.

An important problem that *SpatiHarvest* has to deal with is whether a URL is relevant or not. More specifically, whether a Web page is related to geospatial services (WMSes or WFSes) or not. This decision is critically important both to the *BFC* and *Crawler Controller* components. To deal with this problem, *SpatiHarvest* relies on supervised machine learning. To that end, prior to *Spati-Harvest* execution, a set of pages (relevant and not relevant) has to be collected and properly labeled. The labeled dataset can be used to train a binary classifier and, finally, the trained model can be used by *SpatiHarvest* during its execution. The more accurate the model, the higher precision and recall *SpatiHarvest* can achieve.

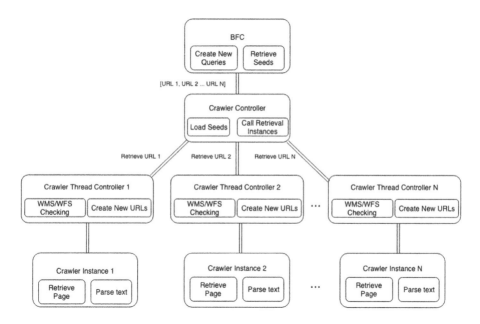

Fig. 2. *SpatiHarvest*'s architecture.

4.2 *SpatiHarvest* Prototype

In order to put *SpatiHarvest*'s architecture to test, we implemented a prototype. Next, we describe our implementation decisions and the prototype's workflow.

Implementation Description. Our prototype was developed using Python 3.6.5 programming language. The choice of such language was due to the following factors: availability of a wide variety of free and open-source libraries, a very active developer community, and a gentle learning curve. Figure 3 presents the UML class diagram of our prototype. The diagram shows the main classes, their methods, and the relationships among them. We discuss the role of the most important classes next.

The *BFC* class is responsible for running the BFC algorithm. In our case, the prototype relies on the Google Search Engine to create an initial set of seeds. This is done via the *ExternalSources* class, which deals with issuing the requests to `google.com` and parsing the responses.

The *CrawlerController* class essentially orchestrates the whole process. It loads seeds via the BFC algorithm, it manages the workers by delegating URLs to instances of the *CrawlerThreadController* class, and it updates the state of the collection process via an instance of the *Frontier* class.

The *CrawlerThreadController* and *Crawler* classes implement the functionalities related to the workers that run in parallel. Our prototype's multi-thread architecture scales vertically, but if computational resources are available, it can be easily extended to scale horizontally. The instances *CrawlerThreadController* are directly managed by the central *CrawlerController* and they are responsible for checking if a page contains WMSes or WFSes and to extract URLs from such a page. The results are then returned to the instance of *CrawlerController*. Each *CrawlerThreadController* is associated with one instance of the *Crawler* class. The latter is responsible for issuing the HTTP request related to a URL (i.e., retrieving the page) and for parsing the content of the response, if any. In terms of libraries, all the text parsing and processing is handled by the *NLTK* library [17], while for the HTTP requests, we relied on *urllib* [18].

The *Frontier* class stores all URLs, whether they belong to pages already visited, pages not visited, or WFS/WMS. Basically, this class is responsible for keeping the consistency of the work that has been already done and the URLs that need to be explored. Also, once a URL has its page collected and parsed, this class is responsible for communicating with an instance of *DAO* class to save the result.

The *DAO* (Data Access Object) class is responsible for providing an interface for database operations. It is useful to isolate all the database logic into a single component, thus reducing the complexity of the code. Underneath the *DAO* class, for the persistence of the retrieved data, the Database Management System (DBMS) NoSQL, MongoDB [20] was used. This choice was due to the demand for a large volume data storage, the flexibility and scalability of MongoDB, and the fact that it is a document-oriented DBMS, which fits perfectly the domain of our problem.

The *SVM* class implements the *SpatiHarvest*'s machine learning capabilities. It has an off-line component for training a SVM (Support Vector Machine) binary classifier prior to any *SpatiHarvest* query and an on-line component, which can be used to decide whether a page is relevant (i.e., related to the WFS and WMS domains) or not during a query processing. In our prototype, the SVM classifier from the *scikit-learn* [19] library, which was trained using the textual content of 100 pages manually labeled. From those, 50 pages were labeled as relevant and 50 as irrelevant. The relevant pages were chosen considering pages containing URLs to WMSes, WFSes, or content closely related to the subject. The irrelevant pages were chosen considering pages with content as different as possible from the relevant pages.

The *Util* class provides a series of useful resources for the proper functioning of the crawler. For instance, the *canCrawlRobots* queries the `robots.txt` of a domain to know if a URL from that domain can, in fact, be crawled by a robot.

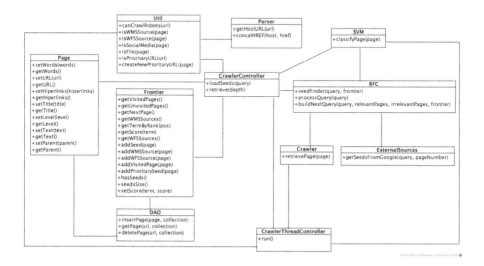

Fig. 3. SpatiHarvest class diagram.

Workflow. Given an input query, the crawler sends it to the *BFC*, which performs the seed retrieval for the beginning of the crawl. Once retrieved and properly stored in the *Frontier*, the *CrawlerController* creates N parallel instances of the *CrawlerThreadController* and assigns to each one the task of retrieving the page associated to each seed or URL not yet visited. This process is repeated successively as long as there are URLs not yet visited in *Frontier*. For each retrieved page, the robot checks if it is a WFS or WMS and if so, the page is added to *Frontier*'s WFS/WMS list. If the page is not a WMS nor a WFS, the *SVM* is used to determine if the page content is or not relevant. If it is relevant, the URLs contained in this page are added to the unseen page queue, if it is irrelevant, those URLs are discarded.

The WFS and WMS identification occurs by combining regular expression matching and keywords search in the page's text. For the WFS verification, first, the crawler looks for the string `service=wfs` in the URL. If there is no occurrence, it is assumed that the page is not a WFS. Otherwise, the HTML tag `<body>` is searched within the page's text. If such a tag is found, it is assumed

---

Algorithm 1: SpatiHarvest

Input: user defined query
Output: list of WMSes and WFSes
if *Frontier has no seed* **then**
    Retrieve seeds using *BFC*
    Add seeds returned to the *Frontier*'s unvisited page list
while *Frontier has seeds* **do**
    Get *Frontier*'s next page
    if *page's level is larger than the maximum allowed search depht* **then**
        continue
    if *URL is priority* **then**
        Create a new priority page
        Add the priority page to the beginning of the unvisited pages list
    Retrieve Web page
    Tokenize all the words contained in the page
    Remove stopwords
    Search for hyperlinks in the page
    Search for hyperlinks in element tags <a>
    Add the retrieved page to the visited pages list and to the database
    if *Retrieved page is a WMS or WFS source* **then**
        Add page to the WMS/WFS list of pages and to the database
    if *Retrived page is classified as relevant by SVM* **then**
        for *each hyperlink in the page* **do**
            if *hyperlink is not a file or a social media link* **then**
                if *link is a prioritary link* **then**
                    Add the page to the beginning of the unvisited pages list
                else
                    Add the page to the end of the unvisited pages list

---

that the page is not a WFS. Otherwise, one last check is performed by looking for the regular expression (`wfs(:|_)(capabilities|featurecollection|getfeature|valuecollection)`). If such a pattern is found in the page's text, the page is finally assumed to be a WFS.

A similar procedure is used for WMS identification. However, during the first step, a search for the string `service=wms` is performed and during the third step, a regular expression is not used. Instead, a search is performed by looking for an occurrence of the strings `wms_capabilities` and `wmt_ms_capabilities`. If any of those are found, there is a confirmation that the page is a WMS.

Algorithm 1 summarizes the *SpatiHarvest*'s workflow. In line 1 the robot checks if there are seeds or unvisited pages loaded in the *Frontier*. If there are none, a new set of seeds is constructed from URLs returned by the *BFC*. Otherwise, those seeds are used for crawling. This verification of unvisited URLs in the *Frontier* becomes useful in the scenario in which the robot, for some reason, has its execution interrupted during a search and then, after some time, it is resumed. In lines 11 through 16, commands are executed in a thread, thus enabling the crawler to retrieve and parse N web pages in parallel. The definition of a *priority* URL, term used in lines 8 and 22, refers to URLs that have `service=wfs` or `service=wms` as substring.

5   Results and Discussion

In order to evaluate our *SpatiHarvest* prototype and compare it to the state of the art, i.e., PolarHub, we conducted several experiments. First, we had also to implement PolarHub, since its code it is not publicly available. During the implementation, we had to make some assumptions, once some details about PolarHub were omitted in their manuscript. Those assumptions were mainly related to the process of deciding whether a given Web page was a WMS or WFS. Both crawlers were run on an instance of the Google Compute Engine, with 26 GB of RAM, 4 vCPUs, and a Premium network level. For each input query, both crawlers were executed simultaneously and the same number of seeds was used in each one.

The experiments were performed with both crawlers and the results are presented in Tables 1, 2 and 3. Table 1 shows results related to the query `brazil wfs services`, Table 2 to query `ogc wms sources`, and Table 3 to the query `ogc web services`. The "URLs Range" column refers to the number of visited URLs. For example, in Table 1, the first row shows the number of WMSes and WFSes identified in the first 10,000 analyzed URLs.

From the numbers, it is possible to observe that *SpatiHarvest* has performed better than Polarhub with regard to WMS and WFS identification. For *SpatiHarvest*, most of the Web services were identified early during the crawling execution and after that, the number dropped significantly. This fact points to the high efficiency of the seed recovery mechanism employed by *SpatiHarvest*, i.e., BFC. Also, during the experiments, it was detected the inability to identify a WMS with the regular expression used by Polarhub, namely `wms(?:t_ms|s)_capabilities`. Such expression did not match with WMS URLs nor with the content of the page returned by the `getcapabilities` functionality of a WMS.

Table 1. Results for query "brazil wfs services" and 89 seeds. SpatiHarvest (Spati.) and PolarHub (Polar.)

URLs range	WMS – *Spati.*	WFS – *Spati.*	WMS – Polar.	WFS – Polar.
1–10000	113	45	0	39
10001–20000	0	2	0	2
20001–30000	0	0	0	5
30001–40000	0	0	0	0
40001–50000	0	0	0	0

Table 2. Results for query "ogc wms sources" and 96 seeds

URLs range	WMS – *Spati.*	WFS – *Spati.*	WMS – Polar.	WFS – Polar.
1–10000	135	48	0	6
10001–20000	1	0	0	0
20001–30000	5	0	0	0
30001–40000	9	0	0	0
40001–50000	6	0	0	0

Table 3. Results for query "ogc web services" and 96 seeds

URLs range	WMS – *Spati.*	WFS – *Spati.*	WMS – Polar.	WFS – Polar.
1–10000	118	33	0	11
10001–20000	1	3	0	2
20001–30000	3	0	0	9
30001–40000	3	0	0	3
40001–50000	4	0	0	4

Figure 4 shows the consolidated results comparing the two crawlers. It was verified that in terms of the number of retrieved Web Map Services, *SpatiHarvest* performed significantly better, even when the query was not directly related to such type of Web Service.

In terms the number of recovered Web Feature Services, *SpatiHarvest* also had results significantly better. However, as previously discussed, our implementation of Polarhub was unable to find Web Map Services due to the problem with its regular expression, which did not match any of the web services found.

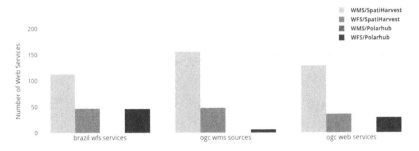

Fig. 4. Comparison between *SpatiHarvest* and PolarHub with regard to the number of identified WFSes and WMSes.

6   Conclusion and Future Work

This paper presented a combination of Web crawling techniques used to conceive *SpatiHarvest*, a contribution to the area of geographic information retrieval. From the experiments that we conducted, it was possible to conclude that our *SpatiHarvest* prototype significantly outperformed the state of the art, PolarHub, with regard to the recovery of Web Map Services and Web Feature Services on the Web. We highlight that our PolarHub implementation was unable to identify WMSes and, to the best of our knowledge, we followed the details presented in the PolarHub's manuscript.

There are three main reasons that explain *SpatiHarvest*'s good performance. First, Bootstrap Focused Crawling enables the *early* discovery of many WMSes and WFSes scattered around the Web. Second, we used fine tuned techniques to decide whether a page was a WFSes or WMSes. Finally, the use of machine learning was a useful pruning technique to filter out pages that were not in the context of geospatial Web services.

Overall, the results show a promising path to *SpatiHarvest*. Despite the fact that our initial work focused on two types of OGC Web services, WFS and WMS, there are a few other types, for which the crawler could be extended to retrieve. There are also other information retrieval techniques that can further improve crawling efficiency, such as hidden web crawling techniques [2,21], where in addition to hyperlinks contained in the pages, forms would also be explored in an attempt to find relevant content. We intend to pursue this research direction in future work.

Acknowledgements. Project partially sponsored by Coordenação de Aperfeiçoamento de Pessoal de Nível Superior (CAPES) and Fundação de Amparo à Pesquisa do Estado de Minas Gerais (Fapemig).

References

1. The Australian government information management office (2019). https://www.finance.gov.au/agimo-archive/better-practice-checklists/docs/BPC17.pdf
2. Raghavan, S., Garcia-Molina, H.: Proceedings of the 27th International Conference on Very Large Data Bases, VLDB 2001, pp. 129–138. Morgan Kaufmann Publishers Inc., San Francisco (2001). http://dl.acm.org/citation.cfm?id=645927.672025
3. Vieira, K., Barbosa, L., da Silva, A.S., Freire, J., Moura, E.: World Wide Web **19**(3), 449–474 (2015). https://doi.org/10.1007/s11280-015-0331-7
4. Bergman, M.K.: The deep web: surfacing hidden value. White paper (2001). https://doi.org/10.3998/3336451.0007.104
5. I. Cyclone Interactive Multimedia Group, I. Cyclone Interactive Multimedia Group. The digital universe of opportunities: rich data and the increasing value of the internet of things (2019). https://www.emc.com/leadership/digital-universe/2014iview/executive-summary.htm
6. Li, W., Wang, S., Bhatia, V.: Comput. Environ. Urban Syst. **59**, 195 (2016). https://doi.org/10.1016/j.compenvurbsys.2016.07.004

7. Baeza-Yates, R.A., Ribeiro-Neto, B.: Modern Information Retrieval. Addison-Wesley Longman Publishing Co., Inc., Boston (1999)
8. Hearst, M.A., Dumais, S.T., Osuna, E., Platt, J., Scholkopf, B.: IEEE Intell. Syst. Their Appl. **13**(4), 18 (1998). https://doi.org/10.1109/5254.708428
9. Chakrabarti, S., van den Berg, M., Dom, B.: Comput. Netw. **31**, 1623 (2000). https://doi.org/10.1016/S1389-1286(99)00052-3
10. Batsakis, S., Petrakis, E.G., Milios, E.: Data Knowl. Eng. **68**(10), 1001 (2009). https://doi.org/10.1016/j.datak.2009.04.002
11. Huang, C.Y., Chang, H.: ISPRS Int. J. Geo-Inf. **5**, 8 (2016). https://doi.org/10.3390/ijgi5080136
12. Broder, A., Mitzenmacher, M.: Internet Math. **1**, 485–509 (2003). https://doi.org/10.1080/15427951.2004.10129096
13. Rodrigues, K., Cristo, M., de Moura, E.S., da Silva, A.: IEEE Trans. Knowl. Data Eng. **27**(08), 2261 (2015). https://doi.org/10.1109/TKDE.2015.2407354
14. Jamali, M., Sayyadi, H., Hariri, B.B., Abolhassani, H.: Web Intelligence, pp. 753–756 (2006). https://doi.org/10.1109/WI.2006.19
15. Debashis, H., Amritesh, K., Lizashree, M.: Int. J. Comput. Appl. **3**, 23–29 (2010). https://doi.org/10.5120/767-1074
16. Pal, A., Tomar, D.S., Shrivastava, S.C.: ArXiv (2009)
17. Natural language toolkit (2019). https://www.nltk.org/
18. urllib - URL handling modules (2019). https://docs.python.org/3/library/urllib.html
19. learn: machine learning in Python - scikit-learn 0.16.1 documentation (2019). https://scikit-learn.org/
20. The most popular database for modern apps (2019). https://www.mongodb.com/
21. Barbosa, L., Freire, J.: Proceedings of the 16th International Conference on World Wide Web, WWW 2007, pp. 441–450. ACM, New York (2007). https://doi.org/10.1145/1242572.1242632
22. Gupta, A., Anand, P.: 2015 International Conference on Futuristic Trends on Computational Analysis and Knowledge Management (ABLAZE), pp. 619–622 (2015). https://doi.org/10.1109/ABLAZE.2015.7154936
23. Cho, J., Garcia-Molina, H.: Proceedings of the 11th International Conference on World Wide Web, WWW 2002, pp. 124–135. ACM, New York (2002). https://doi.org/10.1145/511446.511464
24. Heydon, A., Najork, M.: World Wide Web **2**(4), 219 (1999). https://doi.org/10.1023/A:1019213109274
25. Anandhi, R., Chitra, K.: Int. J. Comput. Appl. **52**, 12 (2012). https://doi.org/10.5120/8172-1485
26. Lopez-Pellicer, F., Florczyk, A., Béjar, R., Muro-Medrano, P., Zarazaga, F.: Online Inf. Rev. **35**, 909–927 (2011). https://doi.org/10.1108/14684521111193193

Targeted Content Distribution in Outdoor Advertising Network by Learning Online User Behaviors

Meng Huang[1], Zhixiang Fang[1,2](✉) [ORCID], and Tao Zhang[3]

[1] State Key Laboratory of Information Engineering in Surveying, Mapping and Remote Sensing, Wuhan University, 129 Luoyu Road, Wuhan 430079, People's Republic of China
zxfang@whu.edu.cn
[2] Collaborative Innovation Center of Geospatial Technology,
Wuhan 430079, People's Republic of China
[3] China Mobile Group Hubei Company Limited, Wuhan 430040, People's Republic of China

Abstract. Digital outdoor advertising has revolutionized the way to display ads by allowing to deliver content dynamically. However, it is still a challenge to show the most suitable ad at the right time and place due to the spatio-temporal dynamics of human preferences. In this study, we attempt to realize smart digital billboards by mining mobile internet usage data, namely, for a given billboard, displaying the most relevant genre of ad at any given time. To achieve this goal, mobile internet usage data is applied as observations to train a state-sharing HMM. The hidden states, i.e., regional interests and preferences are learned, and the relative popularity of the states is then explored spatially and temporally. The regional preference dynamics are revealed by the learned state sequences. The discovered states and sequences of each region can be used to determine the most relevant ad genre to display. The proposed framework was successfully applied to Wuxue, a city in central China, as a case study.

Keywords: Mobile phone internet usage data · HMM · Digital outdoor advertising

1 Introduction

Digital outdoor advertising, as a new form of outdoor advertising, has increased rapidly in recent years. The popularity of digital outdoor advertising is probably due to its flexibility of dynamic displaying multimedia and context related content [1]. Different from static billboards, digital advertising can easily change contents throughout the day according to different passersby. Nowadays, targeted online advertising has been widely applied in the digital world. However, the dynamic display of advertisements in the real-world based on audiences' interests is still at an infant stage due to the difficulties to obtain the user profile in the physical outdoor space. Traditional ways for outdoor advertising have mainly depended on demographic data or census data, which is usually collected at a low temporal frequency and contains little information about people's

© Springer Nature Switzerland AG 2020
S. Di Martino et al. (Eds.): W2GIS 2020, LNCS 12473, pp. 125–134, 2020.
https://doi.org/10.1007/978-3-030-60952-8_13

interests [2]. With the emergence and development of new technology, more and more data sources are available to model crowd behavior and analyze user interests [3, 4]. For example, cameras placed near the digital signage can define the gender, age of the passersby, which are then utilized to select suitable ads to display. However, this method is still insufficient to acquire the customer's interests since only visible attributes can be detected [5]. There are also some dynamic campaigns delivering contextually relevant content based on location, time of day, and weather [6]. Social media data is also mined for targeted dynamic display [2]. But social media data may overrepresent the young adult population [7].

Using mobile apps has become a common activity now, leading to large volumes of spatio-temporal data across the city. Internet usage data can be viewed as the direct indicator of users' interests and has been applied to construct user interest profiles [8]. However, there is little effort to make use of it for targeting outdoor advertising.

In this study, we aim to explore the preference variation of each area from mobile internet usage data by using a Hidden Markov Model (HMM). It is a probabilistic model to simulate the system dynamics as a Markov process with hidden states. HMM contains two essential components: transition distribution and emission distribution. The transition distribution describes how adjacent states are correlated, and the emission distribution describes how the observations are correlated to the hidden states [4].

In this study, the mobile internet usage data are adopted as observations, to explore the hidden states, namely, the regional interests and preferences which govern the app usage by training an HMM-based model. The learned states are used to determine the most relevant ad genre to show. The preference dynamics of each area are revealed by the learned state sequences. Billboard dynamic display can be regarded as the transition between hidden states. However, directly using the classic HMM for modeling regional preference sequence remains a challenging problem: (1) training one HMM for each place often suffers from data scarcity and hard to acquire a reliable model; (2) training a unified HMM for the whole study area will fail to uncover the unique character of each region. Therefore, a state sharing HMM (SSHMM) [9] is applied to address these issues.

Previous studies have proved that regions of the same regional functions usually have similar app usage patterns, e.g., most recreation areas use entertainment apps more frequently [10, 11]. This indicates that different areas can share the same states, which coincides with the design philosophy of SSHMM. SSHMM is different from the classic HMM in two ways: (1) different areas share the same set of latent states; (2) each region has its own transition matrix and initial distribution over different hidden states due to its unique preference. This model has been successfully applied to understand urban dynamics [9]. In this study, it is used to explore regional preference dynamics for targeted outdoor advertising.

2   Methodology

2.1   Generating Online Behavior Observations

As shown in Fig. 1, the first step is to generate model observations from mobile internet usage data. Mobile internet usage data is recorded when a user makes internet access requests. Each record contains an anonymous user ID, the date and timestamp, the website or app visited, ID of the connected mobile phone base station, the amount of consumed data. Table 1 shows the sample records.

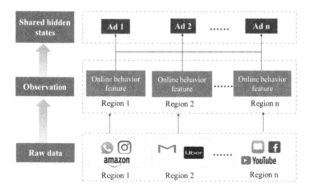

Fig. 1. The framework of this study

Table 1. Sample records of mobile internet usage data

User ID	Date	Time	Base	App	Flow
36**	2015-08-10	08:34	13*	WeChat	0.0107
36**	2015-08-10	09:53	15*	Gaode map	0.0217
⋮	⋮	⋮	⋮	⋮	⋮
36**	2015-08-10	11:30	33*	Taobao	1.32

Since different apps and websites in the same category play similar roles in reflecting the preference of each area, we aggregate them into 8 different categories according to their function. Categories that cannot reflect users' interests are not included, e.g., searching engine. Table 2 shows the number of apps and websites in each category.

Table 2. The 8 app categories used in this study

Category	Number of apps	Category	Number of apps
Social networking	20	Maps and travel	29
Video	41	Finance	36
News	24	Game	28
Online shopping	28	Email and learning	13

When dealing with mobile phone data, the Voronoi polygons generated by mobile phone towers are commonly used subdivisions [12, 13]. In this study, it is used as the basic analysis unit for the study area. A weekday is assumed to contain N ad slots. An ad slot is the minimum time segment to display an ad, which means the content is fixed in a slot. Then based on the timestamp and base station ID of each record, every record is mapped to the corresponding region and timeslot. During $n\text{-}th$ timeslot, the online behavior of each region is represented in the form below:

$$O_{r,n} = \{\hat{w}_{r,n,1}, \hat{w}_{r,n,l}, \ldots \hat{w}_{r,n,8}\} \tag{1}$$

Where $\hat{w}_{r,n,l}$ is the weight of the interest of the category l in the region r during the timeslot n. The most direct way to define the weight is to assign it based on the consumed data of the corresponding app category, but it may result in biased results, because social networking apps are commonly most used, and it will be defined as the most popular category in all the timeslots [14]. In order to eliminate the effects of the data-heavy app categories, the consumed data is normalized by the average data consumption ratio of each app category during each timeslot in the study area. As a result, the weight is calculated as follows:

$$w_{r,n,l} = \frac{r_{r,n,l}}{R_{n,l}} \tag{2}$$

Where $r_{r,n,l}$ is the data consumption ratio of the app category l during timeslot n in the region r. $R_{n,l}$ is the average data consumption ratio of the app category l during the timeslot n. Then it is normalized to ensure all the weights are in the range of 0 to 1.

$$\hat{w}_{r,n,l} = \frac{w_{r,n,l}}{\sum\limits_{l} w_{r,n,l}} \tag{3}$$

After processing each timeslot's data, the online behavior sequences of region r in a day can be defined as the time-ordered sequence below:

$$O_r = [O_{r,1}, O_{r,n} \ldots O_{r,N}] \tag{4}$$

The method is applied to all the regions, and the observation sequence set, as the input of the model, is denoted as follows:

$$O = \{O_1, \ldots, O_r, \ldots, O_R\} \tag{5}$$

2.2 Model Training

In SSHMM, all the input observations O are generated by the shared state set, which is defined as:

$$S = [s_1, s_2, \ldots, s_K] \tag{6}$$

The emission distribution describes how the observations are generated from the hidden states. It is assumed to be Gaussian distribution in this study, so the equation below describes how observations $o_{r,n}$ are generated from the state s_k:

$$p(O_{r,n}|s_n = k) = \prod_{l=1}^{L} \frac{1}{\sqrt{2\pi}\sigma_{k,l}} exp\left(-\frac{(o_{r,n,l} - \mu_{k,l})^2}{2\sigma_{k,l}}\right) \tag{7}$$

Where $\mu_{k,l}$ is the mean of the hidden state s_k. It describes the fundamental feature of the state. $\sigma_{k,l}$ is the variance of the hidden state s_k. The observations generated by s_k are slightly different from the mean and the differences are controlled by the variance. Then SSHMM can be built by $\theta = \{\theta_1, \theta_2, \ldots \theta_r, \ldots \theta_R\}$ with $\theta_r = \{\pi_r, A_r, \mu, \sigma\}$ denoted for the r-th region, where

(1) π_r denotes the initial distribution over K hidden states in the region r, i.e., $\pi_{r,k} = p(s_1 = k)(1 \leq k \leq K)$;
(2) A_r denotes the transition probabilities among the K hidden states in the region r;
(3) μ, σ denotes the mean and variance of observation probability.

To infer the parameters above, the Expectation-Maximization (EM) algorithm [15] is used as the solution. In the E-step, it computes the likelihood function, namely, the probability of the observations represented by the online behaviors for all the regions. Different from the classic HMM, the Q function for all regions is defined as [9]:

$$
\begin{aligned}
Q(\theta, \theta^t) = {} & \sum_{r=1}^{R} \sum_{S} p(S|O_r, \theta_r^t) \ln \pi_{r,k} + \sum_{r=1}^{R} \sum_{S} \sum_{n=1}^{N-1} p(s_{n+1}|s_n) \\
& + \sum_{r=1}^{R} \sum_{S} \sum_{n=1}^{N} p(S|O_r, \theta_r^t) \ln p(O_{r,n}|s_{r,n})
\end{aligned}
\tag{8}
$$

In M-step, it finds a new estimation to maximize the Q function. The EM iteration alternates between performing the E step and the M step until the model tends to converge with the output of parameters $\theta = \{\pi, A, \mu, \sigma\}$.

2.3   Preference Prediction

After model training, the model parameters can be obtained. Then the state sequences can be decoded by the Viterbi algorithm [16]. Based on the last state in the state sequence and the transition probability, the next state can be predicted by maximizing the transition probability. An independent HMM is trained for each region with the same number of states as the baseline. RMSE, the root-mean-square error between the mean value of the predicted next state and the ground truth observation is adopted to evaluate the performance of prediction.

3   Application and Results

3.1   Study Area and Dataset

The proposed framework was applied in Wuxue, a county-level city in China. It consists of 12 sub-districts, among which the Wuxue sub-district has the best base station coverage, and the majority of the residents live there. Hence, we used this area as a demonstration of our model implementation (Fig. 2).

The mobile internet usage data was provided by a major cellular operator in Wuxue. In total, the dataset contains data of 20 days, from the 10th of August 2015 to the 29th of

August 2015. People usually have different preferences during weekends and weekdays. In this study, we mainly focus on the preferences on the weekdays. Hence, 15 weekdays in this dataset are utilized. Users who had at least one mobile internet usage record every day were selected, yielding 8,507 total users for analysis. The data is divided into two parts: the first 12 days for the model training, and the rest for prediction evaluation.

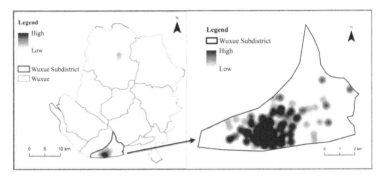

Fig. 2. Location of the study area and spatial kernel density of the distribution of base stations

3.2 Spatio-Temporal Dynamics of Regional Preference

The content of the digital billboard is changed at different frequencies. In this study, we divided a weekday into six timeslots to clearly demonstrate the method. Table 3 shows the time segmentation and the main activity during each timeslot.

Table 3. Advertisement slot

ID	Time period	Main activity
T1	00:00–07:00	Sleep time
T2	07:00–08:00	Morning commute time
T3	08:00–12:00	Morning work time
T4	12:00–14:00	Lunch time
T5	14:00–18:00	Afternoon work time
T6	18:00–24:00	Free time

The error (ε) between the observations and the mean value of the corresponding hidden state are adopted to determine the number of hidden states. From Fig. 3, we can find that when the number of states increases, the error decreases, but the decrease slows down when then state number is bigger than 35. Hence, the state number is set 35 as a trade-off between model complexity and accuracy.

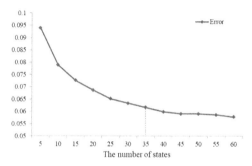

Fig. 3. Error with different number of states

$$\varepsilon = \sqrt{\frac{\sum\limits_{r=1}^{R} \sum\limits_{n=1}^{N} \sum\limits_{l=1}^{L} \left(\mu_{k,l} - o_{r,n,l}\right)}{R \times N \times L}} \tag{9}$$

After training, the 35 states are learned, and the 12 days online observation data can be expressed by the learned states. We mainly focus on the features of the top 15 most frequently appeared states (Fig. 4), since they account for 54% in the state sequences.

Figure 5 shows the dominant states per area for each timeslot. Each color corresponds to a unique state. The state in each region can be used to determine the most relevant ad genre to show at each place and time. It shows that the dominant states vary greatly through time and space. The most frequent state during timeslot 1 is state 24 (k in Fig. 4). This state has a strong preference for video, which indicates people prefer to watch the video at night. The most frequent state during timeslot 2 is state 32 (d in Fig. 4). This state has a higher weight of email and learning, social, map, which indicates that these three categories of apps are frequently used on people's way to work. The most frequent state during timeslot 3 is state 3 (c in Fig. 4). This state has a higher weight of finance, probably because the stock market opens at 9:30 am in China. The most frequent state during timeslot 4 is state 9 (a in Fig. 4), which indicates people prefer to read the news during lunchtime. During timeslot 5, the most frequent state is state 12 (g in Fig. 4). The weight of each category is similar. The most frequent state in the evening is also state 9 (a in Fig. 4).

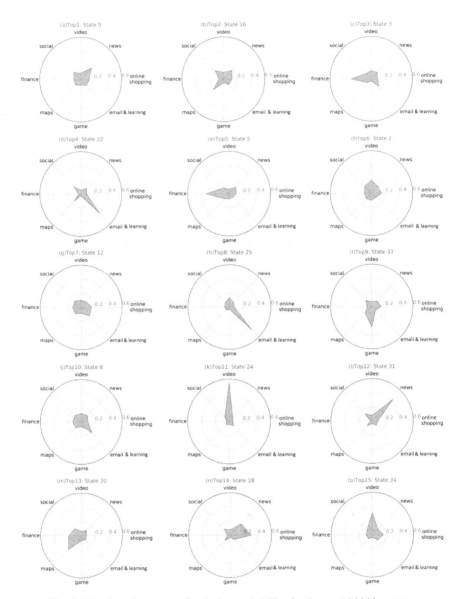

Fig. 4. The Gaussian mean of emission probability for the top 15 hidden states

3.3 Prediction Performance Comparison

The prediction performance of SSHMM and HMM is shown in Fig. 6. Varying the number of states from 2 to 50, SSHMM consistently outperforms the classic HMM. Besides, as the number of states increases, the prediction accuracy of SSHMM is more stable compared with the classic HMM. It indicates that SSHMM can learn reliable models with limited data.

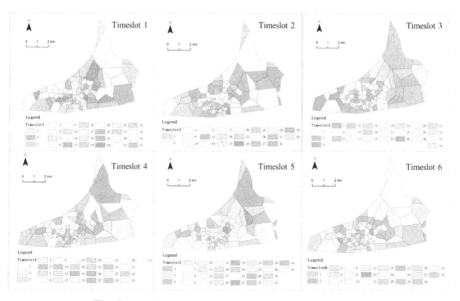

Fig. 5. Dominant states in each area in different timeslots

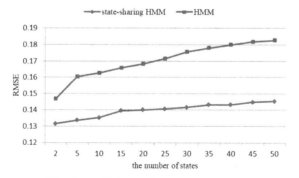

Fig. 6. Prediction performance comparison

4   Conclusion

In this study, we demonstrate how mobile internet usage data can be incorporated with a state sharing HMM for targeted content distribution in the outdoor advertising network. Mobile internet usage data are served as observations. The regional interests and preferences as states are explored by training a SSHMM. The learned states and state sequences provide insights into how interests vary across space and time, which is very informative for targeted outdoor advertising. The discovered states and sequences of each region can be used to determine the most relevant ad genre to display. This framework can also be used for the regional preference prediction.

References

1. Bauer, C., Dohmen, P., Strauss, C.: Interactive digital signage - an innovative service and its future strategies. In: Proceedings - 2011 International Conference on Emerging Intelligent Data and Web Technologies, EIDWT 2011, pp. 137–142. IEEE (2011)
2. Lai, J., Cheng, T., Lansley, G.: Improved targeted outdoor advertising based on geotagged social media data. Ann. GIS **23**, 237–250 (2017). https://doi.org/10.1080/19475683.2017.138 2571
3. Zhang, X.L., Weng, W.G., Yuan, H.Y.: Empirical study of crowd behavior during a real mass event. J. Stat. Mech: Theory Exp. **2012**, P08012 (2012)
4. Zhang, X.L., Weng, W.G., Yuan, H.Y., Chen, J.G.: Empirical study of a unidirectional dense crowd during a real mass event. Phys. A: Stat. Mech. Appl. **392**, 2781–2791 (2013)
5. Okamoto, T., Sasaki, Y., Suzuki, S.: Targeted advertising using BLE beacon. In: 2017 IEEE 6th Global Conference on Consumer Electronics (GCCE), pp. 1–5. IEEE (2017)
6. JCDecaux: Dynamic content (2019). https://www.jcdecaux.com/brands/dynamic-content. Accessed 13 Dec 2019
7. Longley, P.A., Adnan, M., Lansley, G.: The geotemporal demographics of Twitter usage. Environ. Plan. A **47**, 465–484 (2015)
8. Molitor, D., Reichhart, P., Spann, M.: Location-based advertising and contextual mobile targeting. In: 2016 International Conference on Information System, ICIS 2016 (2016)
9. Xia, T., Yu, Y., Xu, F., et al.: Understanding urban dynamics via state-sharing hidden Markov model. In: The World Wide Web Conference, pp. 3363–3369. ACM (2019)
10. Wang, T., Zhang, X., Wang, W.: Spatial-temporal distribution of mobile traffic and base station clustering based on urban function in cellular networks. In: Li, C., Mao, S. (eds.) WiCON 2017. LNICSSITE, vol. 230, pp. 300–312. Springer, Cham (2018). https://doi.org/10.1007/978-3-319-90802-1_26
11. Miao, Q., Qiao, Y., Yang, J.: Research of urban land use and regional functions based on mobile data traffic. In: 2018 IEEE Third International Conference on Data Science in Cyberspace (DSC), pp. 333–338. IEEE (2018)
12. Fang, Z., Yang, X., Xu, Y., et al.: Spatiotemporal model for assessing the stability of urban human convergence and divergence patterns. Int. J. Geogr. Inf. Sci. **31**, 2119–2141 (2017)
13. Xu, Y., Shaw, S.-L., Zhao, Z., et al.: Understanding aggregate human mobility patterns using passive mobile phone location data: a home-based approach. Transportation (Amst) **42**, 625–646 (2015)
14. Huang, M., Fang, Z., Xiong, S., Zhang, T.: Interest-driven outdoor advertising display location selection using mobile phone data. IEEE Access **7**, 30878–30889 (2019). https://doi.org/10.1109/ACCESS.2019.2903277
15. Dempster, A.P., Laird, N.M., Rubin, D.B.: Maximum likelihood from incomplete data via the EM algorithm. J. R. Stat. Soc. Ser. B **39**, 1–22 (1977)
16. Viterbi, A.: Error bounds for convolutional codes and an asymptotically optimum decoding algorithm. IEEE Trans. Inf. Theory **13**, 260–269 (1967)

Understanding Multilingual Correlation of Geo-Tagged Tweets for POI Recommendation

Yuanyuan Wang[1]([⊠])(iD), Panote Siriaraya[2], Mohit Mittal[3], Huaze Xie[3], and Yukiko Kawai[3,4](iD)

[1] Yamaguchi University, 2-16-1 Tokiwadai, Ube, Yamaguchi 755-8611, Japan
y.wang@yamaguchi-u.ac.jp
[2] Kyoto Institute of Technology, Matsugasaki Hashikamicho,
Sakyo-ku, Kyoto 606-8585, Japan
spanote@kit.ac.jp
[3] Kyoto Sangyo University, Motoyama, Kamigamo, Kita-ku, Kyoto 603-8555, Japan
mittal.mohit02@gmail.com, x.huaze2019@outlook.com,
kawai@cc.kyoto-su.ac.jp
[4] Osaka University, 5-1 Mihogaoka, Ibaraki City, Osaka 567-0047, Japan

Abstract. This paper presents a multilingual analysis of Twitter for recommending POIs based on psychographic preferences. People who belong to different countries have different behavioral activities and speak different languages. According to psychographic analysis, for example, people who visit other countries are interested in eating the food of their home country. For this, we aim to clarify psychographic preferences for user behaviors by analyzing geo-tagged tweets based on times, locations, and languages. In this work, we focused on the differences between locations and languages in geo-tagged tweets from European countries. A key feature of the proposed system is the ability to suggest POIs (for tourists) in regions where very few geo-tagged tweets are available in a specific language by using the weighted similarity by others' preferences. Specifically, we first extract languages of tweets, and we identify tweeting countries based on the latitude and longitude of tweet locations. Then, we extract feature words from tweets of a specific language in a specific region by using a *tf-idf* based approach. In this paper, we discuss the POI preferences of different language users based on the linguistic correlation between feature words of tweets in the region.

Keywords: POI recommendation · Multilingual Twitter analysis · Geo-tagged tweets

1 Introduction

In recent years, an increasing number of studies have shown how geo-tagged SNS (Social Networking Service) data from such Twitter could be used to better understand and visualize user behavior [6,9]. In most of these examples,

© Springer Nature Switzerland AG 2020
S. Di Martino et al. (Eds.): W2GIS 2020, LNCS 12473, pp. 135–144, 2020.
https://doi.org/10.1007/978-3-030-60952-8_14

spatio-temporal information used to analyze and predict user behavior. Other types of information, such as contextual information about the users themselves (e.g., age and gender), their emotions and preferences from other users have also been useful in model-based recommendation approaches. However, few studies in this domain have investigated similarities in language usage that could be used to understand the interests and preferences of the general population. Overall, multilingual studies among Twitter users are considered as a niche research topic [14,15,20].

In our previous work on user behavior, we analyzed the differences between data posting locations (times) and locations (times) mentioned in the content [3,10,19]. We also extracted user characteristics (psychographic preferences) on foods by considering user profile languages and commonly used languages [17,18,21]. As a result, these studies have made possible to recommend restaurants for each user in a foreign country based on spatio-temporal information and users' languages. However, the number of POIs (tourist attractions) is less than the number of restaurants, and the methods are not effective for POIs with only a few visits (the number of tweets) for each language [17,18]. Also, people belong to some specific part of the world has some basic psychographic preferences. People, if they visit from some other part of the world, have specific preferences of POIs. Therefore, in this work, we aim to extract psychographic preferences of POIs for each language to calculate the popular degrees of POIs for each language in each foreign country which speaks a different language than the native language. This is done by calculating the correlation between the psychographic preferences of POIs and other languages (as cultural similarities in different countries). Thereby, when a POI with few Italian languages such as "Chelsea Market" in New York, it is possible to newly recommend POIs in French, which has a high correlation with the Italian language. Furthermore, it can also recommend unknown POIs in Seattle with less French languages such as "Pike Place Market". Overall, this work differs from our previous works [17,18] in that instead of just food venues, we recommend general POIs (tourist spots, shops, restaurants, entertainment facilities) for tourism according to the psychographic preferences for POIs where there are very few visits (geo-tagged tweets). In addition, we carry out a separate user evaluation with general POI locations in New York and Boston.

In this paper, we propose a novel Twitter-based POI recommendation method to extract the POI preferences of different language users. This is done by using the linguistic properties of information of Twitter (native languages of users or languages of messages) to infer their POI preferences. To do so, the similarity between different language users in their POI preferences is calculated using correlation metrics and this value is used to infer user preference. Furthermore, we verify the effectiveness of our proposed method through evaluation experiments on extracted POIs with sparse tweets and extracted minor POIs with a low rating in all languages.

The remainder of this paper is structured as follows. In Sect. 2, we discuss previous research that has been carried out related to the spatial and temporal analysis of geo-tagged tweets and multilingual social media analysis. Afterward, in Sect. 3, we describe a POI recommendation method based on linguistic correlation in Twitter. Sect. 4 provides an implemented Twitter-based POI recommender system and discusses our proposed POI extraction method with experiments. Finally, in Sect. 5, we conclude this paper and discuss future works.

2 Related Work

A large number of studies have been carried out on the analysis of the spatial and temporal properties of geo-tagged tweets [4,12]. For example, many studies have shown how various events and social trends could be detected by analyzing content from Twitter. Such trends and events could be detected through the identification of "burst topics" [7] or by utilizing the temporal aspects in tweets [13]. Other studies have shown how an open-domain calendar of significant events could be extracted by Twitter [16]. In addition, geo-tagged tweets have also been used in human mobility research to study and simulate human mobility in the real world both within a local and global context [5].

As Internet adoption increased rapidly due to the growth of the global telecommunications infrastructure, the number of non-English speaking Internet users of social media has grown to become a significant proportion. Thus, one topic of research which has emerged to become a niche area in social network analysis research is the topic of multilingualism and multiculturalism. Most previously mentioned research works do not explicitly consider such aspects when analyzing and predicting user behavior. However, this is becoming more important for research in social media, as a lot of information is spread across different languages and countries. Examples of research in this domain include studies carried out to investigate how multilingual Twitter users mediate between language groups [8,11], and studies which focus more on the technical aspects such as those investigating how language classification models could be developed to determine how many languages are used inside a particular tweet [14]. However, these multilingual studies tend not to incorporate the spatial aspects of tweets (e.g., the tweet user's location). Hence, with a view to future research, it would be valuable to conduct an analysis, at a broader level, about the role of multilingual users on spatial-temporal information.

As such, our work proposes a method in which the multilingual nature of Twitter could be used to suggest food venues for users. While prior systems which were created to recommend items generally rely on contextual information from users (their age and gender, etc.) [2], few studies have investigated how information related to cultural aspects, such as similarities in the language or nationality of users could be used in the prediction of user preferences. Overall, our work is also unique in that it incorporates spatial information together with linguistic properties from a large-scale social media platform in the analysis of user food venue preferences and uses such information to recommend food venues for users. Also, existing recommendation systems are often built around collaborative filtering techniques [1] generally rely on information about the preference from other users to predict user interest. Such a system would generally be difficult to carry out when there is little starting information available. Our approach, however, makes use of publicly available information from social network platforms, allowing recommendations to be generated without requiring an existing internal database of user interactions.

3   Twitter-Based POI Recommendation Based on Linguistic Correlation

In this section, we discuss the linguistic property extraction in any place and the POI recommendation method based on the linguistic correlation in Twitter. In our proposed POI recommender system, we first extract POI names from acquired geo-tagged tweets. Next, we identify tweets in the same language (country) by each posting location (country), and we calculate the similarities of languages based on the correlation coefficient by calculating the *tf* value of each POI. Finally, the system presents the high-scored POIs in the specific region by counting the frequencies of POIs from tweets.

3.1   POI Extraction Based on Locations and Languages

The first step of our method is to identify tweets by a specific period, a specific region, and a specific language. The information about locations, times, native languages, and tweet languages are extracted from geo-tagged tweets Here, we determine the native language if a user selected the language for registration at Twitter, and the tweet language is determined as the language used for the tweeting message. Then, each tweet is considered as a specific language l when the {Native language$_l$} $\vee$ (Tweet language$_l \subseteq \overline{\text{Native language}_l}$).

The second step of our method is to extract POIs which had been visited by each language user, we focused on "I'm at" in tweets (a special type of tweets which users post to denote that they are at a specific location). The geographical coordinates (latitude-longitude) are collected from these tweets. Also, we extract the POIs' names described after "I'm at" in the tweets.

The last step of our method is to recommend POIs in a ranking based on a *tf-idf* based approach in Eq. (1) is used to calculate the rating score of each POI, if a sufficient number of tweets containing a specific POI j is detected in a specific region p for a specific language user (i.e., #tweets > threshold α).

$$\frac{\#\text{tweets of POI}\,j\text{ in the language}\,l_y}{\#\text{tweets of all POIs in the language}\,l_y} \times \log \frac{\text{total}\#\text{languages}\,L}{\#\text{languages in POI}\,j} \quad (1)$$

3.2   POI Extraction Based on Similarity with Languages in Locations with Sparse Tweets

In most cases, there is not a sufficient amount of tweets by a specific language user for meaningful analysis of POI preference for a specific region p (i.e., #tweets < threshold α), it will be a region where visit frequency is low for a specific language

Table 1. The number of collected and analyzed geo-tagged tweets, and the total number of tweets of top-30 POIs in New York and Boston

Language	#Tweets	#I'm @ (%)	New York	Boston
French	906,572	36,938 (4.1%)	1,095	57
Spanish	4,019,581	274,974 (6.8%)	11,652	1,053
Italian	390,957	18,253 (4.7%)	827	185
Portuguese	642,671	26,423 (4.1%)	3,065	331
Total	5,959,781	356,588 (6.0%)	16,639	1,626

l_x. In such cases, we estimate the POI preferences of that language user based on similarities in their POI preferences with other language users from all other regions. More specifically, the similarity $sim(x, y)$ in POI preference between two language users in specific countries (country x and country y) is calculated using the following Eq. (2) based on the cosine similarity.

$$sim(x, y) = \frac{\sum (tf_{\{x,j\}} - \overline{tf_{\{x,j\}}})(tf_{\{y,j\}} - \overline{tf_{\{y,j\}}})}{\sqrt{\sum (tf_{\{x,j\}} - \overline{tf_{\{x,j\}}})^2 \sum (tf_{\{y,j\}} - \overline{tf_{\{y,j\}}})^2}} \tag{2}$$

$$tf_{\{x,j\}} = \frac{\#\text{tweets of POI } j \text{ in the language } l_x}{\#\text{tweets of all POIs in the language } l_x} \tag{3}$$

where $tf_{\{x,j\}}$ denotes the POI frequency of a specific POI j in the language l_x and $tf_{\{y,j\}}$ denotes the POI frequency of the other language l_y. The POI frequency of the POIs in the region (i.e., $tf_{\{x,j\}}$) would be calculated using Eq. (3). Then, the rating scores of POIs in a specific region is calculated using the following Eq. (4).

$$\frac{\sum^D \left(sim(x, y) \times tf_{\{x,j\}} \right)}{\sum^D tf_{\{x,j\}}} \tag{4}$$

where D refers to the total number of languages, the above equation is used to calculate the rating score for the POI j using the language l_x in the region p.

4   Evaluation and Discussion

In this paper, to discuss the feasibility of our proposed POI extraction method in two cities (New York and Boston) of the US, we collected geo-tagged tweets for two years, during 2016/11/12 to 2019/1/7 in four languages (French, Spanish, Italian, Portuguese). Table 1 shows the number of geo-tagged tweets collected in each city and language, as well as the number of POIs, identified from the "I'm at" typed tweets using data about the geographical coordinates (latitude-longitude) of the tweets themselves. Further information about the POIs themselves was obtained using the Swarm API from Foursquare, such as what categories (e.g., genres) are offered in those POIs.

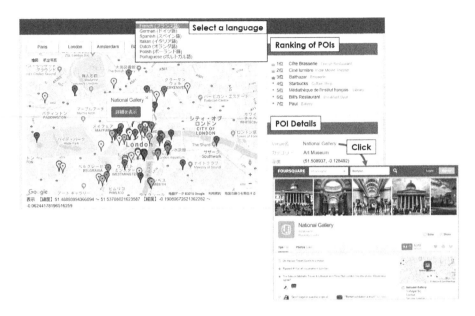

Fig. 1. The user interface of our Twitter-based POI recommender system.

4.1 System Implementation

An interactive map-based web application was developed using PHP, JavaScript, and HTML to visualize the POIs calculated using our approach. Figure 1 shows the user interface of our proposed recommender system. In this system, users first select a language and view its POI preferences in a city. Afterward, markers in different colors (based on different genres) would be shown on the map to represent the recommended POIs for users of that language in the city. On the right side of the map, the POIs ranked by the score would be shown. Users would be able to check the details of each POI by clicking on a marker.

4.2 Diversity Verification of POIs in Languages

Table 2 shows the similarities between languages in New York by our proposed POI extraction method. The highest similarities in other languages to the language l_x are presented in bold. The Spanish language had the highest correlation (0.92) with the Portuguese language, while the French language had the lowest correlation (0.71) with the Portuguese language. On average, the Spanish language had the highest correlation (0.85) with other languages in terms of POI preferences, which the French language had the lowest overall correlation (0.74).

Table 3 shows the similarities between languages in Boston by our proposed POI extraction method. The highest similarities in other languages to the language l_x are presented in bold. The French language had the highest correlation (0.46) with the Spanish language, which was not highly correlated with each

Table 2. The similarities based on POIs in New York of l_x

l_x	FR	ES	IT	PT	Average
FR (French)	1	**0.77**	0.74	0.71	0.74
ES (Spanish)	0.77	1	0.86	**0.92**	0.85
IT (Italian)	0.74	0.86	1	**0.87**	0.82
PT (Portuguese)	0.71	**0.92**	0.87	1	0.83
Average	0.74	0.85	0.82	0.83	0.81

Table 3. The similarity based on POIs in Boston of l_x

l_x	FR	ES	IT	PT	Average
FR (French)	1	**0.46**	−0.03	0.16	0.20
ES (Spanish)	**0.46**	1	0.01	0.44	0.30
IT (Italian)	−0.03	**0.01**	1	−0.09	−0.04
PT (Portuguese)	0.16	**0.44**	−0.09	1	0.17
Average	0.20	0.30	−0.04	0.17	0.16

other. On average, French language and Spanish language had a low overall correlation with other languages in terms of POI preferences, while other languages were not correlated with each other.

From the above, we could confirm the usefulness of our proposed system, and our proposed method has the ability to recommend diverse POIs by the preference of each country based on the analysis of locations and languages.

4.3   User Evaluation Study for POI Extraction

We reported the French language had the highest correlation (0.46) with the Spanish language in Boston in the previous subsection, we selected the French language as the target language since the French language with the smallest number of tweets for POIs in Boston. To evaluate the effectiveness of our proposed POI recommendation approach, we conducted a user study for extracted POIs. Among the native French speakers living in France, 16 and 11 people who visited New York and Boston, respectively, were participated in this user study. Participants were asked to rate the low 10 POIs in a ranking from the top 30 POIs recommended in each city (the top 27 POIs in Boston). The baseline is a ranking for all languages using the general recommendation service, Foursquare rating, we verified user evaluation with the baseline based on Spearman's rank correlation coefficient. In addition, the similarities (Sim) were determined using only New York (NY) in Table 2, and using both New York and Boston (Both).

Table 4 shows the evaluation results by Spearman's rank correlation coefficient. In all the results, POI recommendation by our proposed method was better than the POI recommendation by the Foursquare rating. In addition, Boston

Table 4. The rating of recommended POIs by French speakers

City	Sim	Rating	Proposed	Gain (%)
New York	NY	−0.22	0.22	+44.2%
New York	Both	0.79	0.83	+3.6%
Boston	Both	0	0.55	**+55.0%**
Average	—	0.54	0.14	+39.5%

which has few tweets in the French language, obtained the highest improvement of 55% by our method, which is higher than the Foursquare rating.

From the above, we confirmed the effectiveness of the POI recommendation method based on the proposed linguistic correlation.

4.4   Limitations and Future Work

Overall, there are several limitations to this study. First, the evaluation study was carried out only within the context of French language users and whether the results could be generalized to other language users would need to investigate further. In addition, we focused only on POI recommendation and there is potential for the method proposed in this paper to apply other types.

In our future work, we are looking to develop a method that uses deep learning natural language processing techniques to better identify and match tweets that are more relevant to each corresponding POI to augment the dataset in our system. In addition, we would conduct a more in-depth analysis of the tweets themselves using methods such as sentiment analysis to better reflect the perceived quality of the POIs in our recommendation method.

5   Conclusion

In this paper, we proposed a POI recommender system to extract the POI preferences of different language users by utilizing the linguistic properties of geo-tagged tweets. Information about times, locations, and languages of the geo-tagged tweets could be used to recommend POIs for different language users in different cities, even if the cities themselves contain few tweets of a particular language. Data from different cities were collected and analyzed to recommend POIs based on our proposed method. Afterwards, user evaluation was carried out four language users (French, Spanish, Italian, Portuguese) in two cities (New York and Boston) where users were asked to rate the POIs recommended by our system. The results showed that the linguistic properties of geo-tagged tweets could be useful in recommending general POIs for tourism and not only restaurants as previously identified.

In the future, we plan to apply NLP techniques to identify non geo-tagged tweets. In addition, we would look to expand our recommendation approach to other types (e.g., government service points) and in other geographical areas (e.g., Asia and America, etc.).

Acknowledgment. This work was partially supported by JSPS KAKENHI Grant Numbers JP17K12686, JP19K12240, JP19H04118, and 20H04293.

References

1. Adomavicius, G., Tuzhilin, A.: Toward the next generation of recommender systems: a survey of the state-of-the-art and possible extensions. IEEE Trans. Knowl. Data Eng. **17**(6), 734–749 (2005)
2. Adomavicius, G., Tuzhilin, A.: Context-aware recommender systems. In: Ricci, F., Rokach, L., Shapira, B. (eds.) Recommender Systems Handbook, pp. 191–226. Springer, Boston, MA (2015). https://doi.org/10.1007/978-1-4899-7637-6_6
3. Antoine, E., Jatowt, A., Wakamiya, S., Kawai, Y., Akiyama, T.: Portraying collective spatial attention in Twitter. In: Proceedings of the 21th ACM SIGKDD International Conference on Knowledge Discovery and Data Mining, KDD 2015, pp. 39–48. ACM, New York (2015)
4. Ardon, S., et al.: Spatio-temporal and events based analysis of topic popularity in Twitter. In: Proceedings of the 22nd ACM International Conference on Information and Knowledge Management, CIKM 2013, pp. 219–228. ACM, New York (2013)
5. Birkin, M., Harland, K., Malleson, N., Cross, P., Clarke, M.: An examination of personal mobility patterns in space and time using Twitter. Int. J. Agric. Environ. Inf. Syst. **5**(3), 55–72 (2014)
6. Chen, J., Yang, S., Wang, W., Wang, M.: Social context awareness from taxi traces: mining how human mobility patterns are shaped by bags of poi. In: Adjunct Proceedings of the 2015 ACM International Joint Conference on Pervasive and Ubiquitous Computing and Proceedings of the 2015 ACM International Symposium on Wearable Computers, pp. 97–100. ACM, New York (2015)
7. Dong, G., Yang, W., Zhu, F., Wang, W.: Discovering burst patterns of burst topic in twitter. Comput. Electr. Eng. **58**(C), 551–559 (2017)
8. Eleta, I., Golbeck, J.: Multilingual use of Twitter: social networks at the language frontier. Comput. Hum. Behav. **41**, 424–432 (2014)
9. Hu, T., Song, R., Wang, Y., Xie, X., Luo, J.: Mining shopping patterns for divergent urban regions by incorporating mobility data. In: Proceedings of the 25th ACM International on Conference on Information and Knowledge Management, CIKM 2016, pp. 569–578. ACM, New York (2016)
10. Jatowt, A., Antoine, E., Kawai, Y., Akiyama, T.: Mapping temporal horizons: analysis of collective future and past related attention in Twitter. In: Proceedings of the 24th International Conference on World Wide Web, WWW 2015, pp. 484–494. International World Wide Web Conferences Steering Committee, Republic and Canton of Geneva, Switzerland (2015)
11. Mohd Pozi, M.S., Kawai, Y., Jatowt, A., Akiyama, T.: Sketching linguistic borders: mobility analysis on multilingual microbloggers. In: Proceedings of the 26th International Conference on World Wide Web Companion, WWW 2017, pp. 825–826. Companion, International World Wide Web Conferences Steering Committee, Republic and Canton of Geneva, Switzerland (2017)
12. Mousset, P., Pitarch, Y., Tamine, L.: Studying the spatio-temporal dynamics of small-scale events in Twitter. In: Proceedings of the 29th on Hypertext and Social Media, HT 2018, pp. 73–81. ACM, New York (2018)

13. Nugroho, R., Zhao, W., Yang, J., Paris, C., Nepal, S.: Using time-sensitive interactions to improve topic derivation in Twitter. World Wide Web **20**(1), 61–87 (2016). https://doi.org/10.1007/s11280-016-0417-x
14. Pla, F., Hurtado, L.F.: Language identification of multilingual posts from Twitter: a case study. Knowl. Inf. Syst. **51**(3), 965–989 (2017)
15. Raghavi, K.C., Chinnakotla, M.K., Shrivastava, M.: "Answer ka type kya he?": Learning to classify questions in code-mixed language. In: Proceedings of the 24th International Conference on World Wide Web, WWW 2015 Companion, pp. 853–858. ACM, New York (2015)
16. Ritter, A., Mausam, Etzioni, O., Clark, S.: Open domain event extraction from Twitter. In: Proceedings of the 18th ACM SIGKDD International Conference on Knowledge Discovery and Data Mining, KDD 2012, pp. 1104–1112. ACM, New York (2012)
17. Siriaraya, P., Nakaoka, Y., Wang, Y., Kawai, Y.: A food venue recommender system based on multilingual geo-tagged tweet analysis. In: 2018 IEEE/ACM International Conference on Advances in Social Networks Analysis and Mining (ASONAM), pp. 686–689. IEEE, New York (2018)
18. Siriaraya, P., Wang, Y., Kawai, Y., Nakaoka, Y., Akiyama, T.: Utilizing multilingual social media analysis for food venue recommendation. In: Kaya, M., Birinci, Ş., Kawash, J., Alhajj, R. (eds.) Putting Social Media and Networking Data in Practice for Education, Planning, Prediction and Recommendation. LNSN, pp. 29–49. Springer, Cham (2020). https://doi.org/10.1007/978-3-030-33698-1_3
19. Wakamiya, S., Jatowt, A., Kawai, Y., Akiyama, T.: Analyzing global and pairwise collective spatial attention for geo-social event detection in microblogs. In: Proceedings of the 25th International Conference Companion on World Wide Web, WWW 2016 Companion, pp. 263–266. International World Wide Web Conferences Steering Committee, Republic and Canton of Geneva, Switzerland (2016)
20. Wang, Y., Mohd Pozi, M.S., Siriaraya, P., Kawai, Y., Jatowt, A.: Locations & languages: towards multilingual user movement analysis in social media. In: Proceedings of the 10th ACM Conference on Web Science, WebSci 2018, pp. 261–270. ACM, New York (2018)
21. Wang, Y., et al.: A Twitter-based culture visualization system by analyzing multilingual geo-tagged tweets. In: Dobreva, M., Hinze, A., Žumer, M. (eds.) ICADL 2018. LNCS, vol. 11279, pp. 147–150. Springer, Cham (2018). https://doi.org/10.1007/978-3-030-04257-8_14

Crowdsourcing Trajectory Based Indoor Positioning Multisource Database Construction on Smartphones

Xing Zhang[1], Tao Liu[2(✉)], Qingquan Li[1], and Zhixiang Fang[3]

[1] Shenzhen Key Laboratory of Spatial Information Smart Sensing and Services & Key Laboratory for Geo-Environment Monitoring of Coastal Zone of the National Administration of Surveying, Mapping and GeoInformation, the School of Architecture and Urban Planning, Shenzhen University, Shenzhen 518060, China
xzhang@szu.edu.cn
[2] College of Resources and Environment, Henan University of Economics and Law, Zhengzhou 450001, China
liutao@huel.edu.cn
[3] State Key Laboratory of Information Engineering in Surveying, Mapping, and Remote Sensing, Wuhan University, Wuhan 430072, China
zxfang@whu.edu.cn

Abstract. The bottleneck of fingerprinting-based indoor localization method is the extensive human effort that is required to construct and update the database for indoor positioning. In this paper, we propose a crowdsourcing trajectory based indoor positioning multisource database construction method that can be used to collect fingerprints and construct the radio map by exploiting the trajectories of smartphone users. By integrating multisource information from the smartphone sensors (e.g., camera, accelerometer, and gyroscope), this system can accurately reconstruct the geometry of trajectories. After then, the location of trajectories can be spatially estimated in the reference coordinate system. The experimental results show that the average calibration error of the fingerprints is 0.67 m. A weighted k-nearest neighbor method (without any optimization) and image matching method are used to evaluate the performance of constructed multisource database. The average localization error of RSS based indoor positioning and image based positioning are 3.2 m and 1.2 m respectively.

Keywords: Indoor localization · Crowdsourcing trajectory · Fingerprinting · Smartphone

1 Introduction

Nowadays, indoor localization has become a common issue for various location-based services and applications. A number of technologies have been proposed for indoor localization, based on different principles such as radio frequency (RF) [1], magnetic fields [2], ultra wide band (UWB) [3], ultrasound [4] and so on. Some localization technologies, e.g., ultrasound, can provide accurate localization results. However, these technologies require an extra deployment of localization devices, which restricts their large-scale applications. Therefore, many studies have focused on developing localization scheme that do not rely on extra devices or only use the existing infrastructures, such as Wi-Fi fingerprinting [5–8], visual positioning [9–11] or geomagnetic technology [12]. For example, Wi-Fi fingerprinting can be easily deployed using the existing off-the-shelf mobile devices (e.g., smartphones) and wireless networks (e.g., 802.11 Wi-Fi infrastructures), which makes it very suitable for commercial use.

The collection data is an essential bottleneck for developing localization solutions that independent of extra devices. For example, Wi-Fi fingerprinting-based localization requires a radio map of the whole environment. Visual-based positioning rely on the image database or other semantic information to represent locations. Geomagnetic-based localization depends on a similar map of magnetic field strength of the environment. The collection and updating of required data are quite labor-intensive and time-consuming which prevent these localization solutions from large-scale application. As a wide-spread mobile device, smartphone is suitable for collecting crowdsourcing data, including wireless signals, inertial data, magnetic field, image and so on. While people are walking in different indoor environments, their smartphones can collect the required data continuously at a certain sampling rate. The collected signals are associated with the corresponding sampling points with timestamps. However, the spatial location of sampling points cannot be directly obtained from smartphone built-in sensors in indoor spaces. The geo-tagging of trajectory sampling points is a key issue for crowdsourcing-based localization approaches.

This study proposes to involve visual information in the geo-tagging of sampling points for crowdsourcing based localization scheme. The video frames collected by smartphone camera contain abundant visual information of different spatial location. An advantage of video frame is that it will be little affected by the diversity of smartphone devices. According to our observation, although under different condition of smartphone camera (e.g., resolution ratio, size, etc.), the visual features extracted from these images can accurately reflect the spatial information of sampling points. Consequently, this study attempts to develop a visual-based geo-tagging method for crowdsourcing-based indoor localization. This method can be used to geometrically reconstruct the trajectories (associated with the collected signals) of multiple crowdsourcing users with different smartphone devices. To improve the practicability of this method, prior knowledge, such as an existing database (e.g., Wi-Fi radio map), floorplan of the environment and the initial locations of all crowdsourcing users, is assumed to be unknown. A reference coordinate system (RCS) is defined for the geometry reconstruction of crowdsourced trajectories. In RCS, an initial reference point (IRP) and two axes construct its necessary elements, which can be easily deployed in any indoor environment. An algorithm is also proposed to estimate the spatial location of sampling points from all trajectories. To evaluate the proposed method, the proposed method is used to build a crowdsourced Wi-Fi radio map and evaluate the geo-tagging accuracy of crowdsourced trajectories and the localization accuracy of the built Wi-Fi radio map.

2 Theoretical Framework

We proposed a visual based geo-tagging method for crowdsourcing trajectory based indoor localization framework, and we first illustrate its basic idea. As shown in Fig. 1, some volunteers hold the smartphone (keep the camera forward facing) and walk normally in the indoor area. The video data, inertial data, Wi-Fi and magnetic data can be collected from the smartphone. Firstly, we use the video data and inertial data to recover the trajectory geometry. To improve the location accuracy of trajectory sampling points, the video data and gyroscope data were used for estimating the heading angle. The accelerometer data were used to estimate the step length and step number. Without the initial location and the floorplan, this method can recover the relative locations of trajectory sampling points. Secondly, we calibrate the trajectory sampling points into the indoor reference coordinate system (RCS) by utilizing the initial reference point. Importantly, the calibrated trajectory sampling points can used as supplementary reference point which is aims to estimate more trajectories into RCS. Finally, the calibrated trajectory sampling points with RSS information, magnetic information and image can be directly used for constructing the Wi-Fi fingerprinting database, magnetic fingerprinting database and image database. In summary, this proposed method can construct the database required for indoor positioning by utilizing the crowdsourcing trajectories.

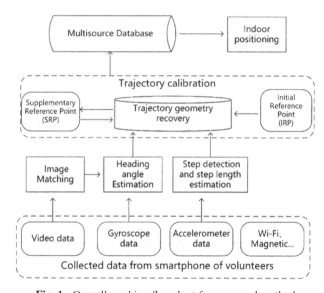

Fig. 1. Overall working flowchart for proposed method

3 Method

In this section, we firstly propose a multi-sensors fusion based approach to reconstruct the geometry of trajectories according to smartphone multi-sensor data, including video frames, accelerometer readings, and gyroscope readings. Then, a trajectory calibration method was proposed to spatially estimate the location of the trajectory geometry based on indoor reference point. Finally, after transferring the crowdsource trajectories into RCS, the trajectory sampling point associate with multisource information e.g. RSS, magnetic, image can construct the multisource database for indoor positioning.

The main idea of trajectory recovery approach is that the heading angle of a sampling point can be represented by the heading angle of the image taken at the sampling point. Therefore, the heading angle change of a sequence of sampling points can be estimated by calculating the heading angle change of the corresponding image sequence extracted from the video frames. In this study, we use an SFM-based method [13, 14] to estimate the heading angle change of sampling points from a trajectory. Firstly, the image matching was implemented on the adjacent two images from image sequence. As referenced in previous work, three constrains were implemented on the scale-invariant feature transform (SIFT) feature [15], which can decrease the mismatching results. The improved image matching result is used for heading angle estimation.

For a pair of images, the homogeneous matching points is used for calculating the fundamental matrix F:

$$\left[u_i', v_i', 1 \right] \cdot F \cdot \begin{bmatrix} u_i \\ v_i \\ 1 \end{bmatrix} = 0 \tag{1}$$

Where $m_i(u_i, v_i, 1)^T$, $m_i'\left(u_i', v_i', 1\right)$ are the homogeneous matching points from image matching result $\left\{ m_i, m_i' | i = 1, 2, \ldots n \right\}$, u_i, v_i are the pixel coordinate from one image and u_i', v_i' are the pixel coordinate from another image. F is an 3×3 order matrix. It is possible to linearly calculate the matrix F if there are enough matched points. After obtaining the fundamental matrix, the essential matrix E should be calculated as:

$$\mathrm{E} = K^T F K \tag{2}$$

Where K is the intrinsic matrix of the smartphone camera, which can be estimated as [16]. The rotation matrix R can be calculated by decomposition of essential matrix E. Importantly, the heading angle of a sampling point can be expressed by rotation matrix:

$$R = \begin{bmatrix} \cos\Delta\theta & 0 & \sin\Delta\theta \\ \sin\Delta\vartheta\sin\Delta\theta & \cos\Delta\vartheta & -\sin\Delta\vartheta\cos\Delta\theta \\ -\cos\Delta\vartheta\sin\Delta\theta & \sin\Delta\vartheta & \cos\Delta\vartheta\cos\Delta\theta \end{bmatrix} \tag{3}$$

Where $\Delta\theta$ is the heading angle change of the sampling point P_t relative to last sampling point P_{t-1}. The schematic diagram of this SFM-based heading angle estimation method is shown in Fig. 2. If the initial sampling point heading angle is θ_0, the heading angle of sampling point P_t can be calculated as:

$$\theta_t = \theta_0 + \sum_{i=1}^{t-1} \Delta\theta_i \tag{4}$$

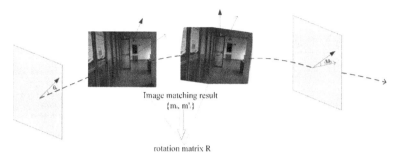

Image matching result
{m., m'.}

rotation matrix R

Fig. 2. The schematic diagram of SFM-based heading angle estimation

The trajectory geometry recovery is to estimate the relative location of each sampling point from a trajectory. The relative location of a sampling point can be calculated as:

$$\begin{cases} x_t = x_{t-1} + L \cdot sin(\theta_{t-1} + \Delta\theta) \\ y_t = y_{t-1} + L \cdot cos(\theta_{t-1} + \Delta\theta) \end{cases} \tag{5}$$

where (x_t, y_t) is the location of sampling point P_t, θ_{t-1} is the heading angle of sampling point P_{t-1}, and $\Delta\theta$ is the heading angle change of P_t that is relative to P_{t-1}. L is the distance between P_t and P_{t-1}. The proposed SFM-based heading angle estimation method can be used to calculate the $\Delta\theta$ of sampling point. However, the accuracy of this heading angle estimation method is depending on the performance of image matching. If the video frames with poor quality e.g. turning with fast speed or turning with blank walls which lead to image blur and few feature points, the proposed method will fail to estimate the heading change of sampling point. In order to solve this problem, the collected inertial data are used to estimate the heading angle, which is independent with visual based heading estimation.

To spatially estimate a trajectory, keypoints from each sample image are matched against those from the reference images (i.e., images of the reference points). If the number of keypoint matches between a reference image and a sample image of a trajectory is higher than a threshold r, the reference point is used to estimate the location of each sampling point (from the same trajectory) by the use of the BA method [17]. Several images previously taken at the IRP are used as its reference images. The main idea of BA is to calculate the 3D location of keypoints and refine the relative location between images by minimizing the projection error of the keypoints and the tracked keypoints on the images. The result of BA is the optimal 3D location of keypoints and the relative pose among the cameras. The spatial relation between a reference point and a sampling point can be represented by the cameras' relative pose calculated by the use of BA. If a sampling point is successfully matched to the IRP, the location of all the sampling points from the trajectory can be estimated in the RCS. The sampling points from the estimated trajectory will used as SRPs for trajectories which do not cross the IRP (Fig. 3).

After transferring the crowdsource trajectories into RCS, the trajectory sampling point associate with multisource information e.g. RSS, magnetic, image can construct the multisource database for indoor positioning. The attributes of the sampling points in RCS are shown in Table 1. Due to the high sampling frequency from trajectory and

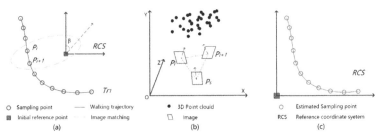

Fig. 3. Trajectory estimation in the RCS. (a) Image matching between the IRP and sampling points. (b) Calculating the relative pose of Ps by the use of the BA method. (c) The estimated trajectory in the RCS.

uniformly distribution in the indoor area, the original sampling points cannot directly be used as Wi-Fi or magnetic based fingerprint. Fortunately, the trajectory sampling points associate with location and image which can directly use for constructing image-based indoor database. To increase the universally of this trajectory sampling points, we design a method for constructing wireless signal (Wi-Fi and magnetic) based indoor database.

Table 1. The attributes of the sampling points.

Sampling point ID	Trajectory ID	Coordinates	RSS	Magnetic strength	Image
P1	Tr_1	(X_1, Y_1)	$\{(rss_1, ap_1), (rss_2, ap_2)...\}$	(x1, y1, z1)	I1
P2	Tr_2	(X_2, Y_2)	$\{(rss_1, ap_1), (rss_2, ap_2)...\}$	(x2, y2, z2)	I2
P3	Tr_3	(X_3, Y_3)	$\{(rss_1, ap_1), (rss_2, ap_2)...\}$	(x3, y3, z3)	I3

4   Evaluation

The estimation of recovered trajectory geometry is based on SFM-based heading angle and PDR. To quantify the error of recovered trajectories, we design four routes with known initial locations and all the trajectories were uniformly sampled with marker associate with coordinates to obtain the ground-truth points. During the experiment, the participants held the smartphone forward facing at a fixed posture and walked along each route at normal pace. The inertial and video data of smartphone were recorded continuously from the trajectory beginning to the end. Another participant responsible for recording the times when participants walked past each marker. Each route was repeated 10 times for calculating the location of sampling points. Figure 4 shows the four routes and the real scene in this experiment. As shown in Table 2, the average location error of sampling points from four trajectories is around 0.5 m, the standard deviation of location error is about 0.4 m. Thus, the proposed trajectory geometry recovery method can obtain accurate estimation results.

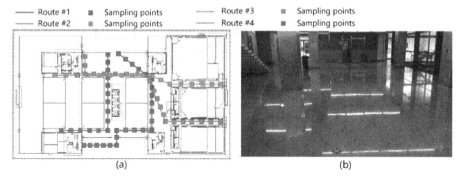

Fig. 4. Four routes associate with marker to evaluate the trajectory geometry recover method. (a) is the ground-truth location; (b) is one of real scene in experiment

Table 2. Error of trajectory geometry recovery.

Route Number	Length/m	Average error/m	Standard deviation/m
Route#1	82	0.55	0.4
Route#2	65	0.47	0.35
Route#3	85	0.58	0.42
Route#4	59	0.41	0.38

To further evaluate the performance of this geometry recovery approach, two other methods were also used for comparison: the gyroscope-based method and the SFM-based method. Gyroscope based method is only use the gyroscope data from smartphone to calculate the heading angle and restore walking trajectory. SFM-based method is to use video frames and SFM method (not integrate inertial data) to estimate the heading angle. The shape discrepancy metric (SDM) was used to verify the accuracy of these method, it defined as the Euclidean distance between a sampling point and its corresponding ground-truth point. Figure 5 shows the cumulative distribution function (CDF) of the SDM for 40 trajectories using the three methods. Clearly, the SDM error of the gyroscope-based method is much higher than that of the other methods. For the SFM-based method, the maximum SDM error is about 3 m; the 80-percentile SDM error is around 2 m; and the mean SDM error is about 1.1 m. This indicates that visual information can help to improve the estimation performance of the trajectory recovery. Moreover, the SDM error can be further reduced by the use of the visual/inertial fusion based method: the maximum SDM error is about 1.5 m; the 80-percentile SDM error is around 1 m; and the mean SDM error is about 0.53 m. These results demonstrate that the fusion of the visual and inertial information helps to overcome the drawbacks of the single-source based methods, e.g., drift error from the gyroscope or matching failure of SFM. The visual/inertial fusion based approach can be used to recover trajectories with different geometries. Furthermore, the experimental trajectories covered wide spaces in the study area. The results demonstrate that this approach can perform well in a wide indoor space, which is a difficult environment for the commonly used gyroscope-based methods.

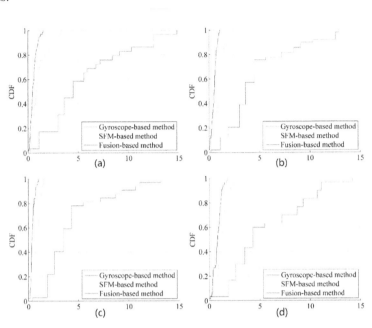

Fig. 5. The CDF of annotation errors. (a) Route#1. (b) Route#2. (c) Route#3. (d) Route#4.

After calibrating the sampling points of trajectory, the multisource database such as RSS, magnetic and image can be constructed based on proposed method. To evaluate the quality of the constructed RSS database, the online localization was conducted based on the weighted k-nearest neighbor method at the 66 test points (the centers of the 66 grids). Each of these grids was covered by more than five trajectories. The reason for selecting these grids was to test whether the increase of the crowdsourced trajectories helped to improve the quality of the constructed radio map. For comparison, a site survey process was also conducted based on the same mesh grids. The localization error was calculated as follows:

$$Err_i = \sqrt{\left(x_i^r + x_i^e\right)^2 + \left(y_i^r + y_i^e\right)^2} \tag{6}$$

Where Err_i is the location error of point i, $\left(x_i^r, y_i^r\right)$ is the actual physical location of point i, and $\left(x_i^e, y_i^e\right)$ is the estimated physical location of point i.

The localization results of site survey method and this proposed method are shown in Fig. 6. The bar of R0 represents the localization error of the site survey method. The bars of R1 to R5 represent the localization error of this proposed method. R1 refers to the set of fingerprints for which the RSS came from one trajectory. Similarly, R2, R3, R4, and R5 refer to the sets of fingerprints for which the RSS came from two, three, four, and five trajectories, respectively. As can be seen in Table 3, the localization errors of the site survey based method range from 0 to 4.9 m, and the average error is 2.6 m. The average error of R0 is 4.3 m, which is higher than that of R0. With the increase of the trajectories (from R1 to R5), the average error gradually decreases and reaches 3.2 m when the RSS of the fingerprints is extracted from more than four trajectories. This indicates that the quality of the constructed database is at the same level as the site survey based database when there is sufficient crowdsourced data. The proposed system can greatly reduce the human labor needed for indoor database construction. Moreover, it performs well in wide indoor spaces, which increases the potential for applying this system to large indoor environments such as shopping malls, underground parking garages, or supermarkets.

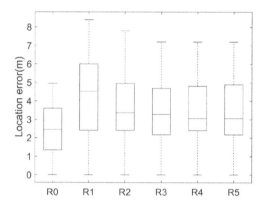

Fig. 6. Localization results of the two methods

Table 3. location error of different method, where R0 refers to site survey method, R1 refers to trajectory method from one trajectory, R2, R3, R4, and R5 refer to the sets of fingerprints for which the RSS came from two, three, four, and five trajectories.

Database	R0	R1	R2	R3	R4	R5
Max error	4.9	8.4	7.8	7.2	7.2	7.2
Average error	2.6	4.3	3.7	3.4	3.2	3.2
Error std	1.3	2.2	1.9	1.8	1.8	1.8

To evaluate the quality of the constructed image database, a SURF based image matching method was used [18] to calculate the similarity among query images and reference images. The average location error of image matching based method is 1.2 m, the accurate matching rate is 94%. Compare with other image matching based method [19], this method achieves a relative higher accuracy and matching rate. The reason is that the sampling points collected by this method are at a high spatial sampling rate along a trajectory, which contribute to the constructed image datasets with also high spatial resolution. It helps to improve the performance of image matching and reduce the location error of image matching based positioning. Compare with [20], this method does not need laser backpack to construct a 3D model. The time and equipment cost of this method are relatively low, which is important to crowdsourcing-based data collection and indoor localization.

5 Conclusion

The collection and update of the indoor positioning database is an unavoidable bottleneck for fingerprinting-based indoor localization. The traditional site survey is quite labor-intensive and time-consuming, which limits the commercial and industrial application of fingerprinting-based indoor localization. In this paper, we proposed a crowdsourcing trajectory based indoor positioning multisource database construction method. Firstly, this method can recover the trajectory geometry by integrating inertial and visual data from smartphone. Then, the crowdsourcing trajectories can be spatially estimated to the indoor reference coordinate system by using initial reference point and bundle adjustment algorithm. Finally, the multisource database would be constructed in different types of indoor spaces, such as corridors, rooms, or wide spaces. Prior knowledge, such as floorplans or initial locations of crowdsourced trajectories, is not required as initial inputs, which will facilitate the application of this system. The experimental results demonstrated that the integration of visual and inertial information can significantly improve the performance of trajectory recovery and estimation. The average location error of the RSS based positioning and image based positioning are 3.2 m and 1.2 m respectively.

References

1. Wang, A.Y., Wang, L.: Research on indoor localization algorithm based on WIFI signal fingerprinting and INS. In: International Conference on Intelligent Transportation, Big Data & Smart City (ICITBS). Xiamen, China, January 2018, pp. 206–209 (2018)
2. Namkyoung, L., Ahn, S., Han, D.: AMID: accurate magnetic indoor localization using deep learning. Sensors **18**, 1598 (2018)
3. Chen, P., Kuang, Ye., Chen, X.: A UWB/Improved PDR integration algorithm applied to dynamic indoor positioning for pedestrians. Sensors **17**, 2065 (2017)
4. Díaz, E., Pérez, M.C., Gualda, D., Villadangos, J.M., Ureña, J., García, J.J.: Ultrasonic indoor positioning for smart environments: a mobile application. In: IEEE 4th Experiment@ International Conference, Faro, Algarve, Portugal, June 2017, pp. 280–285 (2017)
5. Bahl, P., Padmanabhan. V.N.: RADAR: an in-building RF-based user location and tracking system. In: Proceedings of IEEE INFOCOM 2000. Nineteenth Annual Joint Conference of the IEEE Computer and Communications Societies. Tel Aviv, Israel, March 2000, pp. 775–784 (2000)
6. Youssef, M., Ashok, A.: The horus WLAN location determination system. In: Proceedings of the 3rd International Conference on Mobile Systems, Applications, and Services. ACM, 2005, pp. 205–218 (2005)
7. He, S., Chan, S.: Wi-Fi fingerprint-based indoor positioning: recent advances and comparisons. IEEE Commun. Surv. Tutor. **18**, 466–490 (2017)

8. Tao, L., Xing, Z., Qingquan, L., Zhixiang, F.: A visual-based approach for indoor radio map construction using smartphones. Sensors **17**, 1790 (2017)
9. THOMPSONS. IndoorAtlas. http://www.Indooratlas.com
10. Ravi, N., Shankar, P., Frankel, A., Iftode, L.: Indoor localization using camera phones. In: IEEE Workshop on Mobile Computing Systems & Applications. Orcas Island, WA, USA, August 2006, pp. 1–7
11. Chen, Y., Chen, R., Liu, M., Xiao, A., Wu, D., Zhao, S.: Indoor visual positioning aided by cnn-based image retrieval: training-free, 3D modeling-free. Sensors **18**, 2692 (2018)
12. Ruotsalainen, L., Kuusniemi, H., Bhuiyan, M.Z.H., Chen, L., Chen, R.: A two-dimensional pedestrian navigation solution aided with a visual gyroscope and a visual odometer. GPS Solut. **17**, 575–586 (2013)
13. Argyriou, V., Del Rincón, J.M., Villarini, B.: Structure from Motion. Wiley, Chichester (2015)
14. Wu, C.: Towards linear-time incremental structure from motion. In: 2013 International Conference on 3DV-Conference. IEEE Computer Society (2013)
15. Yao, L., Feng, H., Zhu, Y.: An architecture of optimised SIFT feature detection for an FPGA implementation of an image matcher. In: International Conference on Field-Programmable Technology, 2009. FPT 2009. IEEE (2010)
16. Li, Y., Zhuang, Y., Lan, H., et al.: A Hybrid WiFi/Magnetic Matching/PDR approach for indoor navigation with smartphone sensors. IEEE Commun. Lett. **20**(1), 169–172 (2015)
17. Torr, P.H.S., Zisserman, A.: Vision Algorithms: Theory and Practice. Springer, Heidelberg (2000). 1883, https://doi.org/10.1007/3-540-44480-7
18. Bay, H., Tuytelaars, T., Gool, L.V.: SURF: speeded up robust features. In: Proceedings of the 9th European Conference on Computer Vision - Volume Part I. Springer, Heidelberg (2006)
19. Ravi, N., Shankar, P., Frankel, A., et al.: Indoor localization using camera phones. In: IEEE Workshop on Mobile Computing Systems & Applications. IEEE (2006)
20. Liang, J.Z., Corso, N., Turner, E., et al.: Image based localization in indoor environments. In: Fourth International Conference on Computing for Geospatial Research & Application. IEEE (2013)

A Social-Spatial Data Approach for Analyzing the Migrant Caravan Phenomenon

Roberto Zagal-Flores[1], Miguel Felix Mata[1], and Christophe Claramunt[2(✉)]

[1] Instituto Politécnico Nacional (IPN), Mexico City, Mexico
{rzagalf,mmatar}@ipn.mx
[2] Naval Academy Research Institute, Brest, France
claramunt@ecole-navale.fr

Abstract. The migrant caravan that recently came out from Central towards North America generated polarized opinions in online social networks. The objective of this paper is to explore the social spatial-temporal trends that emerge from this migrant caravan phenomenon, and based on a combination of social media and newspaper opinions and reports, together with additional socio-economic data. The framework combines text data mining, text clustering, sentiment analysis and spatiotemporal data exploration. The study reveals significant ethnic polarization and ideological patterns but noticeable regional differences in rural and urban areas. The experimental study shows that our approach provides a valuable experimental framework to study emerging regional phenomena as they appear from social media.

Keywords: Geosocial analysis · Spatial data science · Local events analytics

1 Introduction

The migrant caravan, that came out from South America to the USA in 2018, generated massive comments on online social networks. People's opinions as they appear in social networks provide valuable means for a better comprehension of the social-psychology characteristics of the concerned population. Social networks facilitate the expression and spread of polarized views and ideologies and have been the subject of a growing body of research across disciplines [1, 3]. Nowadays, immigration phenomena are sensitive issues that are often discussed in social networks, and that can even lead to the expression of hostility towards minorities, migrants, and people of foreign origin [7].

The migrant caravan routes started from El-Salvador and Guatemala to the Mexico-US border in Tijuana [12]. Our aim is to focus on the migrant caravan routes when analyzing the perception of this phenomenon, and when extracting and interpreting people's opinions. This closely relates to interdisciplinary research, whose objective is to develop appropriate natural language and statistical analysis. A common objective is to correlate expectations between different regional and social groups concerning to given phenomenon [11].

© Springer Nature Switzerland AG 2020
S. Di Martino et al. (Eds.): W2GIS 2020, LNCS 12473, pp. 156–165, 2020.
https://doi.org/10.1007/978-3-030-60952-8_16

The semantic analysis and text clustering combined with statistical analysis, it has been applied to assess the combination of personal traits and engagement with harmful online behaviors [2]. For instance, the Bayesian classifier developed to classify racist tweets in [9]. Geographical differences, as reflected by location-based data, in the expression of opinions about social events have been recently conducted in [5].

The research developed in this paper is based on a dual approach that combines a layered machine learning architecture that explores iteratively key terms, with a geographical analysis of the trends that appear in order to generate polarized maps. The objective is to discover insights as they appear from a social network and a regional phenomenon. The methodology applies an architecture layered based on machine learning techniques, such as text clustering, sentiment analysis, location detection, crossing-data, and spatiotemporal data exploration [11]. These techniques are applied to more than 8,000 tweets and 300 memes closely related to the migration caravan for two months in 2018. The rest of the paper is organized as follows. Section 2 briefly introduces the methodology principles and data mining approach applied to the socio-economic dataset. Section 3 reports on the experimental results. Finally, Sect. 4 concludes the paper and outlines future work.

2 Methodology

2.1 Data Integration

The extraction process from Twitter is based on terms and hashtags related to migrant topics. About 8,000 tweets were recollected from November to December 2018. Tweet locations were collected when available. Initially, hashtags from news media were also considered, thus facilitating the extraction process. The initial hashtags identified by exploring the Twitter accounts of the leading newspapers in Mexico were as follows: #CaravanaDeMigrantes (Migrant Caravan) 2587 tweets, #CaravanaMigrante (Migrant Caravan) 3740 tweets, and #NoMasInmigrantes (No more migrants) 200 tweets. Last, still using the mentioned hashtags for the same period time, 300 text memes were recollected manually due to Facebook restrictions.

2.2 Discovering Social Insights

The purpose is to detect insights in tweets and memes using a robust text mining process that combines text clustering, topic modeling, and sentiment analysis. The discovered patterns in space and time are also correlated to the population profile data and news media. Our approach is inspired by [13], where the authors proposed an LDA model for detecting features in complex events about natural disasters.

Identifying Topics: Text clustering and topic modeling reveal the internal structure of documents and interrelations. Clustering searches for similarity between documents to categorize them so that similar documents are in similar clusters. Topic modeling essentially integrates soft clustering with dimension reduction. Documents are associated with a number of latent topics, which correspond to both document clusters and compact representations identified from a corpus. The Latent Dirichlet Allocation (LDA) method can also be used in order to model the topic distribution of a new document [6].

Accordingly, a K-means clustering algorithm [4, 15] has been applied, together with integration of a LDA algorithm [14]. The complete process uses the following steps: (1) Running a K-means algorithm, the input is a dataset of filtered tweets and text memes. Data is filtered by time and location. The initial values for the K clusters were selected for data exploration purposes in order to identify key topics and then to further trigger different K values that range from 10 to 50 groups. The output gives a set of similar text groups. (2) Then, the output K-means is the input of the LDA process which estimates the inter-entity similarity between clusters of tweets, assuming that elements of one cluster should be similar to one another and dissimilar to the elements of other clusters [11]. The result gives an understandable representation of documents that is useful for analyzing the document themes [6]. The output is a set of topics that could describe the collective opinion reflected in tweets and text memes. This process was written in a Python script using machine learning and NLP capabilities of Gensim and SciKit Learn libraries. Gensim is the main Python library for Paragraph Vector embeddings and well documented, it has been used in [16] for detecting alerts during flooding events with useful results. (3) Finally, the discovered topics are used to create specific queries for cross-analyzing news media and tweets data, the idea being to discover insights and events related to the phenomena.

Sentiment Analysis: Tweets categorized in positive, negative, and neutral sentiment categories by a Semantic-Bayesian classifier based on our previous work [10]. First we run a Bayesian classifier with a sentiment-based reference dataset, which has more than 250 negative and positive comments. Texts were preprocessed by lemmatization process, ignoring stop words and popular regional slang term. Secondly, the a second step performs a word search using a Web Ontology Language (OWL) semantic hierarchy that is an adaptation of WordNet-Affect. This OWL hierarchy contains negative and positive labeled concepts organized by sentiment categories (e.g., sadness, annoyance, anger, joyfulness, happiness). Our previous work has been extended by a third step, where a combination of the two aforementioned ones is performed, using the OWL ontology, especially when the Bayesian classifier cannot derive sentiment classes. This option produces better results in comparison with these previous steps applied in a separated way.

Location Detection: This process is applied for tweets without location reference, the purpose is to analyze speeches by geographical region. The process uses a Web Service provided by Meaning Cloud's API to identify names places in texts. It is improved by geographic gazetteers that contain locations on migrant caravan routes. The steps are as follows: (1) meaning Cloud's API return the list of name places and the texts that belong to "GeoPoliticalEntity" location category, 2) Words are searched in gazetteers to confirm a location reference, 3) If a text contains location references identified by step 1 and 2, then the geographical coordinates are recovered.

2.3   Crossing Data for Retrieval Correlated Geo-events

This stage allows an analysis of topics, polarization, and data correlation considering the spatial distribution in order to discover patterns and events that should favor a better understanding of the migrant caravan phenomenon. This task is composed of two steps: 1) spatial data cube modeling for polarization distribution, and 2) data crossing in news media and population profile data.

Spatial Data Cube Modeling: A data cube is a collection of aggregate values classified according to several properties of interest (i.e., dimensions) [17]. In order to explore the data and discover patterns in space and time, we designed a series of spatial data cubes for social data considering the following dimensions: {[datetime], [likes], [shares], [Topic], [Location reference], [Polarity], [Cluster_Size]}. "[Date-Time]" corresponds to the tweet creation time, "[Latitude]" and "[Longitude]" are related to coordinates where a text was extracted, "[Cluster_Size]" describes the number of documents related to a specific topic, [Location reference] is a name of place discovered in the previous location process.

Spatiotemporal distribution of opinion polarization is obtained by a data cube with {[datetime], [Location reference]} dimensions, and by applying a count operation on the [(Polarity)] dimension that contains how many times a sentiment (positive, negative or neutral) appears in a specific region and time. Moreover, in order to visualize what are the topics and polarity that describe migrant social texts, we propose cloud words plots, and data cube with {[Location reference], [Topic], [Polarity], [likes]} and the sum operation on the [Topic] dimension, that is, the size of the clusters that emerged in different locations or with a similar name. This visualization strategy is inspired from a previous work [8], where authors explored the political polarization in the United States.

Crossing Data: The objective is to discover associated local events in news media, using topics, location (i.e., result of the location detection phase) and tweet attributes to design web requests in Google news RSS Feed. The input parameters of the request query contain{topic name, month, year, place's name}, for example: {EEUU agents, November, 2018, Tijuana}. The outputs give a list of regional events in newspaper data that could be related to the discovered patterns and geographic polarization in migrant caravans. Besides, the population profile layer makes the difference between rural zones, commercial districts, and urban regions. This allows understanding correlations between population profiles and migrant-based opinions through spatial intersection operations. The National Institute of Statistics and Geography (INEGI, Mexico) provided the population data, time coverage is from 2011 to 2017.

3   Experimental Results

We developed a series of bottom-up analyses from the city to the country granularity levels. It appears that most of the opinions expressed regarding the migration caravan were negative comments (e.g., more than 800 negative comments generated on November 25, 2018; negative comments - red sliced; positive comments - green sliced; neutral comments - yellow sliced). Figure 1 shows the polarization of these patterns using chart pies over a few selected locations and tweets extracted from 10 km buffers. The second trend that appears is given by the geographical distribution of opinions still on the migration caravan (Fig. 2). After applying a term clustering process, contrasted trends appear with a negative distribution in Tijuana (North of Mexico), Mexico City and Chiapas (South of Mexico) while other regions exhibit rather positive opinions. Some terms exhibit contrasted polarization (e.g., Tijuana, and Donald Trump).

On the basis of the most frequent terms (Fig. 2) (i.e., violence, Tijuana, migrants) identified on November 25 from people's opinion, a second analysis was performed using news media and social data. The media findings indicate more than five news reported about migrants who tried to evade a Mexican police blockade in Tijuana. Social media shows strong negative reactions, for example "Mexicans organize themselves to expel migrants - Protests in Tijuana https://t.co/kvocWJO5KJ" that can be considered as xenophobia according to [7].

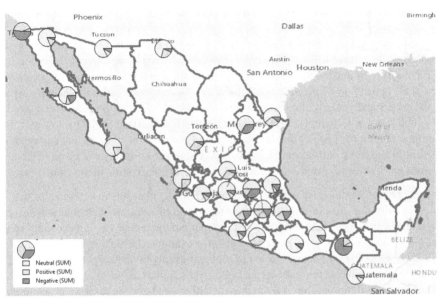

Fig. 1. Polarization map about the migrant caravan opinions in Mexico (Color figure online)

Baja California, was one of the major destinations of the migrant caravans, while being characterized by semi-rural areas (orange squares), as well as residential zones, commercial districts and industrial parks (blue polygons) (Fig. 3). More than 1000 texts were identified as referenced to the Tijuana Mexican border (Fig. 3, a buffer of 10 km shows the regions were tweets were extracted) (red semi-circumference). About 47% of the texts expressed were classified as negative opinions (red disk) and 30% as positive comments (green disk), the rest are neutral comments (yellow disk). It also appears that the opinions expressed were by relatively young people as the average resident population age in Mexico, between 18 and 35 years according to data government (including Tijuana, Mexico City and Chiapas).

Fig. 2. Negative versus positive migrant topic word clouds (Color figure online)

A close analysis of the terms exhibited in Tijuana reveals some contrasted patterns. Topics such as "migrant shelter" (albergue), refuge (refugio), and asylum (asilo) stand out. A close analysis of the data gathered across tweets, pictures and news reports show a lot of comments on shelter conditions, citizen complaints and asylums requests. Another worthwhile topic is "children" (niños), that is used for citizens to report possible police misconduct: "On the border with Tijuana, the US is launching tear gas against#CaravanaMigrante, against children https://t.co/OZhvLdoxw0".

Mexico City is not only a migrant crossing point, but it also concentrates a diverse population and activities in residential zones, commercial districts, industrial parks (green polygons), and a few rural regions. The analysis reveals a wide variety of opinions. Figure 4 shows the patterns that appear at the North, that is, mostly neutral (yellow disks) or rather positive (green disks) or balanced opinions (green and red disks), while at the South with rural and residential zones, 35% of opinions appear as negative.

Overall, Mexico City appears as a place of high opinion contrasts. For example, migrant topics such as "humanitarian crisis" (crisis humanitarias) are likely to express favorable opinions in social media and newspapers (Fig. 5). On the other hand strong terms such as "hate" (odio), "migrant" (migrante) and "hondureño" mostly present in tweets and memes are indeed related to citizen complaints, racist comments. For example: "I would rather feed a dog than a Honduran person" https://t.co/TEWiCZh9ql.

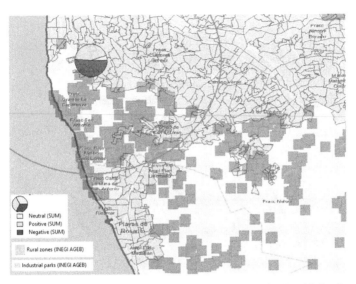

Fig. 3. Opinions expressed in Tijuana semi-rural region (border of USA) (Color figure online)

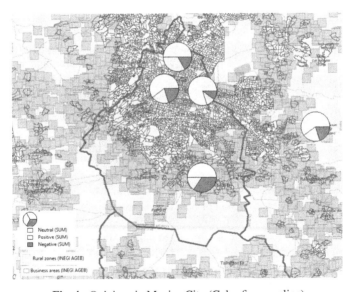

Fig. 4. Opinions in Mexico City (Color figure online)

On the other hand, Chiapas, a border region South of Mexico close to Guatemala, a rural Mexican state, expresses much more balanced opinions. Figure 6 reveals neutral opinions in Tapachula (Guatemala border South of Chiapas), but much more negative comments in San Cristobal de las Casas (the central Capital of Chiapas).

Fig. 5. Cloud migrant topics in Mexico City (CDMX)

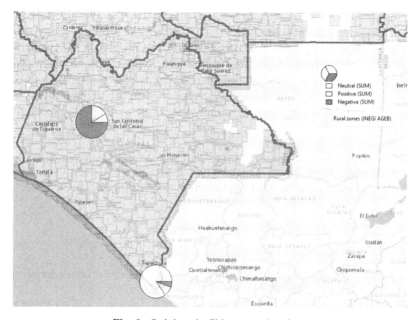

Fig. 6. Opinions in Chiapas rural regions

Last, the topics shown in the region of Chiapas, such "#DisplacedIndigenousPeople" hashtag (indígenas desplazados) show a sort of transfer of opinion as the migrant caravan is closely related to another important social issue in Mexico, the one of indigenous people displacements. For instance "For me, It would be better to support the #DisplacedIndigenousPeople and not the #MigrantCaravan #WeAreChiapas https://t.co/tDo yY1pZkV". The same trend is reflected in local news report: "Displaced indigenous people face police in Chiapas http://mile.io/2RhfAa8#MigrantCaravan".

4   Conclusions

The experimental research developed in this paper introduces an applied methodology to explore the social spatial-temporal trends associated to the migrant caravan phenomenon considered as a regional phenomenon, and in combining opinions expressed in reports on social media, newspaper and memes' text. The framework is based on a combination of text mining and machine learning algorithms and spatio-temporal data analysis exploration. From the experiments done, the findings show strong polarization in the opinions related to the migrant caravan, although there are noticeable differences in rural, urban areas and population characteristics. Although only 8000 tweets were recollected about migrant caravan (considering the limitations of the twitter API for data extraction), this sample represents relatively well the Mexican population in terms of age (18 and 35 years old), the sample also belong to important cities of the migrant caravan routes such Mexico City, Chiapas and Tijuana. The overall framework can be still extended and improved. For instance, the text mining process can be enriched using hierarchical clustering algorithms and deep learning. This might help to find additional topics in the extracted tweets aspects as well as identifying additional text locations. Overall future work will be oriented towards refinements of the social analysis of the findings that emerge, as well as continuous monitoring of the phenomenon across social and newspaper media and application of the whole framework to other social media phenomena.

Acknowledgments. The authors want to thank God, SIP (Secretaría de Investigación y Posgrado), IPN (Instituto Politécnico Nacional), COFAA (Comisión de Operación y Fomento a las Actividades Académicas del IPN), ESCOM and UPIITA IPN for their support.

References

1. Lee, S., Xenos, M.: Social distraction? Social media use and political knowledge in two US Presidential elections. Comput. Hum. Behav. **90**, 18–25 (2019)
2. Bogolyubova, O., Panicheva, P., Tikhonov, R., Ivanov, V., Ledovaya, Y.: Dark personalities on Facebook: Harmful online behaviors and language. Comput. Hum. Behav. **78**, 151–159 (2018). https://doi.org/10.1016/j.chb.2017.09.032
3. Bode, L.: Political news in the news feed: Learning politics from social media. Mass Commun. Soc. **19**(1), 24–48 (2016)
4. Srivastava, A.N., Sahami, M.: Text Mining: Classification, Clustering, and Applications. CRC Press (2009). 328 pages
5. Witanto, J. N., Lim, H., Atiquzzaman, M.: Smart government framework with geo-crowdsourcing and social media analysis. Future Generation Computer Systems. Hruby, F.: From third-person to first-person cartographies with immersive virtual environments. Proc. Int. Cartogr. Assoc., 2, 44 (2019). https://doi.org/10.5194/ica-proc-2-44-2019
6. Aggarwal, C.C, Zhai, C.: Mining Text Data. Springer Editorial
7. Council of Europe (1997). https://rm.coe.int/CoERMPublicCommonSearchServices/DisplayDCTMContent?documentId=0900001680505d5b. Accessed 16 Feb 2017
8. Rouse, R.: Political Polarization in the US, 14 September 2017. https://public.tableau.com/en-us/s/gallery/political-polarization-us

9. Huang, Q., Zhang, L., Cheng, Y., Li, P., Li, W.: Detecting tweets against blacks. Adv. Synth. Catal. **360**(17), 3266–3270 (2018). https://doi.org/10.1002/adsc.201800642
10. Miranda, C.A., Rodriguez, R.C., Zagal-Flores, R.: Arquitectura Web para análisis de sentimientos en Facebook con enfoque semántico. Res. Comput. Sci. **75**, 59–69
11. Bi, T., Liang, P., Tang, A., Yang, C.: A systematic mapping study on text analysis techniques in software architecture. J. Syst. Softw. **144**, 533–558 (2018). https://doi.org/10.1016/j.jss.2018.07.055
12. Cranley, E.: This is insider, news digital media, October 2018. https://www.thisisinsider.com/where-the-caravan-is-map-coming-to-us-border-mexico-2018-10
13. Chen, X., Zhou, X., Sellis, T., Li, X.: Social event detection with retweeting behavior correlation. Expert Syst. Appl. **114**, 516–523 (2018). https://doi.org/10.1016/j.eswa.2018.08.022
14. Agarwal, A., Singh, R., Toshniwal, D.: Geospatial sentiment analysis using Twitter data for UK-EU referendum. J. Inf. Optim. Sci. **39**(1), 303–317 (2018)
15. Srivastava, A.N., Sahami, M.: Text Mining: Classification, Clustering, and Applications. CRC Press (2009)
16. Barker, J.L.P., Macleod, C.J.A.: Development of a national-scale real-time Twitter data mining pipeline for social geodata on the potential impacts of flooding on communities. In: Environmental Modelling & Software, vol. 115, pp. 213–227 (2019). ISSN 1364-8152
17. Giuseppe, S., Leonardo, T.: Statistical dissemination systems and the web. In: Handbook of Research on Public Information Technology, pp. 578–591. IGI Global (2008)

A Mobile Trusted Path System Based on Social Network Data

Félix Mata[1]([✉]), Roberto Zagal-Flores[1], Jacobo González León[1], and Christophe Claramunt[2]

[1] Instituto Politécnico Nacional (IPN), Mexico City, Mexico
{mmatar,rzagalf}@ipn.mx
[2] Naval Academy Research Institute, Brest, France
claramunt@ecole-navale.fr

Abstract. Social networks provide rich data sources for analyzing people journeys in urban environments. This paper introduces a trusted path system that helps users to find their routes based in low crime rate and theft reports. These data are obtained geo-tagged tweets and an official database. Recommended paths are computed automatically from these data sources by a complementary application of social mining techniques, Bayes algorithm and an adaptation of the Dijkstra algorithm. This system can be also used to identify the probability that an event occurs in specific locations and times. A proof of concept of the system is illustrated through two example scenarios.

Keywords: Trusted paths · Recommender systems · Outdoor navigation

1 Introduction

Every day millions of people pass through the streets of the city of Mexico, taking some specific routes to reach their destination using either public or private transportations or even walking. Although well-known routes are most often taken, new routes might be experimented especially for newcomers. This can be considered as a non-straightforward task, generally addressed by computerized assisted-systems that search for either shortest or fastest paths [6]. However, it is has been recognized that more semantics should be integrated in the search process. In particular, safety might be a critical parameter to take into account, especially in overcrowded large cities such as the city of Mexico. The level of trust closely associated to a specific route should reflect a low level of crime rate, or in other words a low probability to be a victim of a theft or crime. The question that arises is how to evaluate the level of trust of a given route?

2 Methodology

A series of recommender systems have been so far developed for route planning in urban systems. Several approaches such as as CrowdPlanner [2] and DroidOppPathFinder [3]

© Springer Nature Switzerland AG 2020
S. Di Martino et al. (Eds.): W2GIS 2020, LNCS 12473, pp. 166–170, 2020.
https://doi.org/10.1007/978-3-030-60952-8_17

take into account real-time road network conditions. Recent years have witnesses a large range of experimental applications derived from social network and crowdsourcing data [1, 4]. For example, in [4] a crowd-based routing system is developed to improve the recommendation quality, but the approach is only based on volunteer-based inputs. However, and to the best of our knowledge, these trusted route search systems haven't been applied in combination with urban statistical data.

The methodology used is first based on the application of a data mining process. The social data source is a tweets database (i.e., 320,000 tweets obtained in a period of three months in the city of Mexico) filtered by previously defined semantic characteristics (e.g., those that contain the word or words "theft", "violence", among others). The processes successively applied to the tweets repository are as follows: (A) **filtering** in three sub-steps: data cleaning, unification and data cleaning and where the objective is to tokenize tweets (B) **semantic matching** where an ontology and Bayes algorithm are applied to identify the most relevant tweets associated to a location using a previous designed ontology of crime/thefts reports and (C) **data assessment and presentation** on a map interface. A sample of two different records as extracted from the Twitter dataset at different levels of spatial granularity is shown below:

1.- @diariodfmx: crime report in public transport, with a fire arm, Iztapalapa
2.- @Dangeraway: Theft to students in Ecatepe, street 105 and 106 @alertaEcatepec

A tweet is potentially associated to a record of the official database according to its relevance. Relevance is measured by contextualizing the record and the tweet. This means that each semantic element of the tweet and record is first identified, and then searched into the ontology of crime and theft in order to find matching terms, synonyms or related concepts as suggested in [7]. For example, the first tweet above is generic in location but rich regarding the description of the event (the contrary for the 2nd one). This process generates a semantic classification matched to the official database categorization, and where each tweet is described according to a few categories either Booleans or descriptors (e.g., theft, crime, violence etc.). Location is derived by neighborhood, street name or relation (e.g., near, far, etc.). Time is computed based on explicit data or temporal adverbs (e.g., now, afternoon, etc.). Tweets with other attributes that cannot be classified are kept as additional parameters for the application of the Bayes algorithm in stage (C). The Bayes algorithm is applied at the stage (C). It classifies the incoming data generated by the filtering process, as explained in the previous section, and generates some predictive spatio-temporal patterns according to some given indicators and rules. The Bayes algorithm searches for some hidden patterns in a given dataset D similar to some substructures. In other words the objective is to search for the probability for places in a given route to have some theft or offenses when passing through. These probabilities are based on the following combinations:

P (coordinate = [x, y] | event); P (day or Time | event); that results in: P(c|x) = Posterior probability: the probability that an event occurs based on past events, and where *P(x) = Total Probability:* denotes the number of times some given attributes appear in the events (e.g., theft). *P(c|x) = Conditional probability:* denotes the number of times a target appears in each attribute (e.g., car), considering the total number of times the target appears in all attributes. *P(c) = A priori probability:* denotes the number of times a target appears in all events. The probability that an event occurs on a given day at a given place is then derived as follows:

$$P(c|x) = P(x) * P(x|c)/P(x) \qquad (1)$$

All the events, attributes and locations identified are then classified by the Bayes algorithms. Places are identified and diagnosed as trusted or untrusted and normalized in a range [0, 1] from the lowest to the highest probability.

ClassifiedEvent ('12/05/2014','3', Monday,1,2255928567);
[{"evento":"Theft","idCoordenada":"30339694","lat":19.4038744,"lon":-
99.150226,"probability":"0.0009295401"},{diagnosed weight: 0.89}"

A trusted path is derived by an adaptation of the Dijkstra algorithm where nodes in the network are assigned a dynamic weight (i.e., values obtained from the Bayes algorithm application) that reflects the probability of having or not an event such as: theft/crime. These weights can be also relaxed by spatial and temporal values: date (day, month, year), location (point or zone) and time (hour or period). At the interface level, a user can configure the route search process, that is, (1) if the route is generated using either social, official or a hybrid data source. The search process also makes a difference between (2) different transportation means. The user can also define a temporal period considered for both data sources.

3 Preliminary Results

The experimental developments have been tested using Android phones version 3.0 to 4.0 by a panel of 10 students over a period of 20 days. The results and the examples shown below were made with a data block of 2541 events derived from the tweets dataset over a period of one month. From this dataset, a frequency table is generated, indicating the number of times each attribute appears given the events theft, crime, etc. The values of the frequency are used as weights when deriving a trusted route. These values are all assigned to corresponding nodes in the network. In addition, the probability that an event happens at some location and date is computed and stored in a similarity table. This also allows comparison of crime probabilities at different places. This sort of summarization can provide a support to evaluate the probability of an event to occur at a specific location and/or particular time (e.g., evaluation of the probability of a crime or theft to occur in avenue Eje Central on Mondays?). Then, by applying Eq. 1, the probability of an event to occur at a given place and time can be derived, for instance on Monday at 9:00 am in Iztapalapa, we have:

$$P(x) = P(Monday) = 220/2149; \; P(x|c) = P(Monday|theft) = 18/218$$

$$P(c) = P(theft) = 218/2149;$$

$$P(c|x) = P(Monday|theft) = P(Monday) * P(theft|Monday)/P(Monday) = 0.0818$$

An analysis sequence can be summarized as follows: first the probabilities of some classified events (theft, crime, etc.) are computed. Next, the selected combinations (e.g., a theft at a particular location and time) are derived. Finally, this gives the probabilities for all events to occur for each particular value of domain. For instance, this denotes the probability that a theft occurs at a given location on Friday, or a crime to occur Tuesday at noon. The example shown above is based on locations derived from a given place, but the system can also iterate all possible combinations close to this location, days and hours checking multiplying events. Figure 2 and 3 show a path between two points, where circles represents untrusted points for a user walking, and shortest path (Fig. 1) and trusted route (Fig. 2).

Fig. 1. Shortest path (walking)

Fig. 2. Trusted path (in car)

Similarly, Fig. 3 shows a usual shortest route generation by a driver user. Alternatively, the user configures the system to consider events that occurred in car only, for deriving a trusted shortest path (Fig. 4).

Fig. 3. Shortest path

Fig. 4. Trusted path

4 Conclusions

This experimental work developed in this paper introduces a preliminary prototype of a trusted route navigation guide. The system developed recommends trusted paths based on an integration of complementary information derived from a large tweet repository and crime database reports of the city of Mexico. The framework combines an analysis of a Twitter dataset with a probabilistic approach based on a Bayes algorithm whose objective is to categorize the levels of crime/thefts in the urban network. Such an analysis supports the derivation of trusted paths. The advantage of the approach is that it combines long-term statistical data with almost real information from citizens acting in the city. The system developed has been tested in the city of Mexico. Ongoing works are oriented to additional validations and comparison with other recommender systems.

Acknowledgments. The authors want to thank God, SIP (Secretaría de Investigación y Posgrado), IPN (Instituto Politécnico Nacional), COFAA (Comisión de Operación y Fomento a las Actividades Académicas del IPN), ESCOM and UPIITA IPN for their support.

References

1. Adomavicius, G., Tuzhilin, A.: Toward the next generation of recommender systems: a survey of the state-of-the-art and possible extensions. IEEE Trans. Knowl. Data Eng. **17**(6), 734–749 (2005)
2. Su, H., Zheng, K.,Huang, J., Jeung, H., Chen, L., Zhou, X.: Crowdplanner: a crowd-based route recommendation system. In: IEEE 30th International Conference on Data Engineering (ICDE), pp. 1144–1155, March 2014
3. Arnaboldi, V., Conti, M., Delmastro, F., Minutiello, G., Ricci, L.: Droidopppath_nder: a context and social-aware path recommender system based on opportunistic sensing. In: 14th IEEE International Symposium and Workshops on World of Wireless, Mobile and Multimedia Networks (WoWMoM), pp. 1–3, June 2013
4. Nagarajan, M., Gomadam, K., Sheth, A.P., Ranabahu, A., Mutharaju, R., Jadhav, A.: Spatio-temporal-thematic analysis of citizen sensor data: Challenges and experiences. In: 10th International Conference on Web Information Systems Engineering, pp. 539–1553 (2009)
5. Ceikut, V., Jensen, C.: Routing service quality? Local driver behavior versus routing services. In: 4th International Conference on Mobile Data Management, pp. 97–106 (2013)
6. Zhang, H., Xu, Y., Wen, X.: Optimal shortest path set problem in undirected graphs. J. Comb. Optim. **29**(3), 511–530 (2014). https://doi.org/10.1007/s10878-014-9766-5
7. Bezzazi, E.H.: Building an ontology that helps identify criminal law articles that apply to a cybercrime case. In: ICSOFT (PL/DPS/KE/MUSE), pp. 179–185 (2007)

Modeling Individual Daily Social Activities from Travel Survey Data

Dan Zou and Qiuping Li[✉]

Center of Integrated Geographic Information Analysis, School of Geography and Planning,
Sun Yat-Sen University, Guangzhou 510275, China
liqp3@mail.sysu.edu.cn

Abstract. Inferring activity types from the massive human-tracking data is of great importance for the understanding of human daily activity patterns in the cities. Researchers have investigated various methods to infer activity types automatically, however, the recognition accuracy of social activity types (such as shopping, schooling, transportation, recreation, and entertainment) are not satisfactory. This research proposes a machine-learning-based method to model individual daily social activities from travel survey data. Using Guangzhou as an example, we extract 21 dimensional spatial and temporal attributes to construct the random forest (RF) method to identify and validate social activities at the individual level. The experiment result shows the recognition accuracy of our approach is 75%. The effects of different factors on social activity participation are also investigated. The proposed approach can help us better understand human behaviors and daily activities, and also provide valuable insights for land use and traffic management planning and other applications.

Keywords: Social activities · Travel survey data · Random forest model · Guangzhou

1 Introduction

Individuals' travel trajectories collected by GPS devices or mobile phones provide new opportunities to understand human mobility patterns. They have sufficient travel location information but generally lack activity information. If activity information, especially social activity types, such as shopping, schooling, and eating out, can be inferred from individuals' daily travel characteristics as well as the urban built environment, then the trajectory data will be significantly enriched. It will enhance our understanding of human behaviors and activities and can provide valuable insights for land use and traffic management planning and other applications [1, 2].

The present studies on activity identification mainly focused on three types of activities: at home, at work, and other activities or social activities [3, 4]. Social activities are not performed as regularly as those at home or work, so it is not easy to identify them accurately. The identification of social activities is important for us to understand the mobility of urban residents and their use of urban space. Recently, some studies have started to combine the travel survey data and machine learning methods to implement

© Springer Nature Switzerland AG 2020
S. Di Martino et al. (Eds.): W2GIS 2020, LNCS 12473, pp. 171–177, 2020.
https://doi.org/10.1007/978-3-030-60952-8_18

social activity identification and verification at the individual level [5, 6]. For instance, Diao et al. developed an activity detection model with travel dairy surveys and applied the modeling results to mobile phone traces to extract potential activity information [5]. Zhu used the travel survey data of Singapore to identify the types of travel activities in the card swiping data from buses [6]. These studies have made a good exploration in the activity identification with travel survey data. However, the identification accuracy of social activities is not satisfactory. Besides, the impact of modeling methods and the selected spatial and temporal attributes on the inference accuracy are still needed to be investigated.

In this paper, we extracted 21 dimensional spatial and temporal attributes that related to individuals' social activities from the travel survey data of Guangzhou in 2017. Then, the RF method was used to identify the social activities at the individual level and compared with other machine learning methods. The overall recognition accuracy of the proposed approach is about 75%. Besides, the importance of the selected spatial and temporal attributes that influence individuals' social activity choices were analyzed.

2   Study Area and Data Source

Guangzhou is located at the mouth of the Pearl River Delta region in south China, which is one of the largest cities in China, with a population of 14 million in 2017 [7]. The research data in this paper is the household travel diary survey of urban residents in Guangzhou in 2017. With the stratified sampling method, thirteen representative communities in six core administrative districts (Yuexiu, Haizhu, Tianhe, Baiyun, Liwan, and Panyu districts) were selected for the household travel survey. The locations of thirteen survey communities are shown in Fig. 1. The activity types of each individual on a working day are recorded every 15 min. The locations of these activities are also recorded synchronously. After data cleaning, 630 social activities on a working day from 997 individuals are selected for analysis. Besides, built environment information is obtained by Point of Interest (POI) data to explore its influences on social activity choice. The POI dataset was crawled from Gaode Map. It contains 14 primary categories, such as hotels, restaurants, companies, finance and business buildings, education facilities, and tourist attractions.

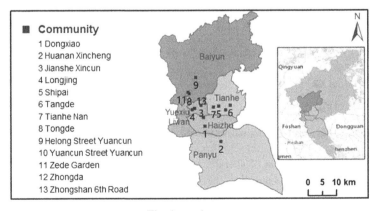

Fig. 1. study area

3   Methodology

3.1   RF Method

RF method uses multiple decision trees to train samples of classification [8]. Although there are many machine learning algorithms, the RF method has witnessed an extensive application for many research fields and achieved success for its ability to reduce biases [9]. It combines the bootstrapping and the random feature selection. In the bootstrapping, every single tree grows with a different training sample, which is randomly selected from the training dataset with replacement. Through sampling with replacement, some observations might appear more than once, while others will be left out in the bootstrap sample, known as out-of-bag (OOB) observations [10].

In this study, we use the RF method to learn the relationship between the explanatory variables and the social activity types. Then, social activity types can be inferred from trajectories. To validate the method, we employed both accuracy and Kappa coefficient to test samples of our data. Finally, the Gini impurity index [9] is utilized to identify significant explanatory variables that contribute to social activities.

3.2   Social Activities Categories

We referred to the existing literature on the classification of activity types [5, 6], and aggregated travelers' social activities of our research into four types. They are personal business (5%) (i.e., going to banks or hospitals), entertainment (27%) (i.e., shopping, doing outdoor exercise, or going to friends' parties), eating out (50%), and other activities (17%) (i.e., buying food or picking up family members).

3.3   Explanatory Variables

Individuals' daily activities are related to activity locations, built environment, weather, and temporal attributes [5, 6]. Therefore, the explanatory variables of the RF method in this research include the location, built environment, weather, and temporal attributes, which are presented in Table 1. Then, for simplicity, we divided the research area into fine grids to obtain the aggregated values of those explanatory variables. The size of the grids is set as 500 m * 500 m [1, 5].

Table 1. Descriptive statistics of variables in the model.

Variable		Min	Max	Mean	Std.dev.
Location attributes	F1: Distance to the nearest bus stop (m)	8.04	1177.23	221.38	104.18
	F2: Distance to the nearest MTR station (m)	52.36	7043.76	665.79	838.42
	F3: Distance to home (m)	0.00	26432.06	4684.70	4443.30
Built environment	F4:Population density (people/500 m^2)	0.00	418.08	268.00	104.41
	F5-F16: Number of POIs in each category	0.00	186.00	7.74	13.70
Weather	F17: Precipitation	0.00	28.60	0.61	2.23
	F18: Temperature	24.86	36.72	30.05	2.66
Temporal attributes	F19: Start time	6.25	22.00	14.57	3.98
	F20: End time	7.33	24.00	15.74	4.27
	F21: Activity duration	0.03	5.50	1.16	0.77

4   Results

4.1   Results of the RF Model

An RF implementation in the Scikit-learn package of Python (https://pypi.org/project/sci kit-learn/) is used to train the RF model. To determine the optimal parameters, we tested the influence of the various number of decision trees on the model performance. The sample data were randomly stratified into two subsets: 75% of the total sample is used as training data, and the remaining 25% dataset is used as testing data. Predicted accuracy and OOB error were used to evaluate the performance. As Fig. 2 shows, when there are 160 decision trees in the RF model, the optimal model performance can be obtained with accuracy at almost 75% and OOB error at 0.28. Figure 3 shows the confusion matrix, and its Kappa coefficient is 0.6229. The activity of eating out is classified with the highest accuracy (85%), followed by other activities (71%). The accuracy of entertainment and personal business activities is lower, which is 62% and 50%, respectively.

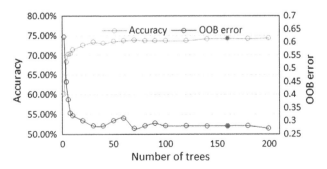

Fig. 2. RF performance with the number of trees

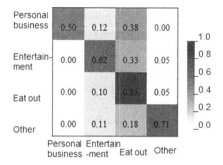

Fig. 3. Confusion matrix for the RF model

To analyze the importance of variables that affect individuals' social activities, we used the Gini index [9] to measure the performance of variables when trees split in the training process. Figure 4 shows the importance of all 21 explanatory variables (F1 to F21 in Table 2). The effects of the explanatory variable are different. The importance of all the temporal attributes (F19 to F21) is high, especially the duration of activity (F21), which means the temporal attributes are good signals of social activity types. The location attributes (F1 to F3) are also essential to model social activities, especially the distance to home (F3). The built environment and the weather variables are less important than the abovementioned two variable sets.

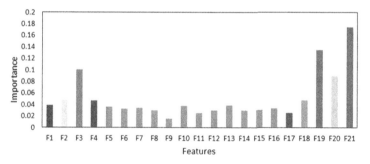

Fig. 4. Variable importance in RF model

4.2   Model Comparison and Accuracy Assessment

A comparison is also made among RF method, support vector machine (SVM) [11], adaptive boosting (AdaBoost) [12], and multinomial logit (MNL) model [13], to find the method that performs the best in social activities inference. The same training data and testing data are used to train these models and evaluate their performance. Table 2 shows the overall prediction accuracy. It can be observed that RF performs the best, with total accuracy as high as 75%. The next-best performer is the AdaBoost model, followed by MNL. The reason is the testing dataset has a small personal business activity sample, and MNL and SVM are sensitive to minority samples and thus yield inaccurate results. Therefore, we conclude that RF is a promising method to improve the inference accuracy of individuals' social activities.

Table 2. Performance of different models

Predicted category	RF	AdaBoost	SVM	MNL
Personal business	50%	50%	11%	3%
Entertainment	62%	54%	59%	64%
Eat out	85%	78%	87%	78%
Other	71%	68%	28%	58%
Total	75%	69%	65%	67%

5 Conclusion

This research proposes a machine-learning method to model individual social activities from travel survey data. Taking Guangzhou as an example, the overall inference accuracy of our approach is 75%. The effects of different factors on the social activity types are also investigated, which can help us better understand human behaviors and activities, and also provide valuable insights for land use and traffic management planning and other applications. Our future work is to apply the modeling results to mobile phone location data to extract the potential social activity information.

Acknowledgments. This study was supported by the National Natural Science Foundation of China (Grant # 41971345) and the Guangdong Basic and Applied Basic Research Foundation (Grant # 2020A1515010695).

References

1. Tu, W., et al.: Coupling mobile phone and social media data: a new approach to understanding urban functions and diurnal patterns. Int. J. Geographical Inf. Sci. **31**(12), 2331–2358 (2017)
2. Rasouli, S., Timmermans, H.: Activity-based models of travel demand: promises, progress and prospects. Int. J. Urban Sci. **18**(1), 31–60 (2014)
3. Furletti, B., et al.: Inferring human activities from GPS tracks. In: Proceedings of the 2nd ACM SIGKDD International Workshop on Urban Computing. ACM (2013)
4. Huang, W., Li, S.: An approach for understanding human activity patterns with the motivations behind. Int. J. Geographical Inf. Sci. **33**(2), 385–407 (2019)
5. Diao, M., et al.: Inferring individual daily activities from mobile phone traces: a Boston example. Environ. Plan. Planning Des. **43**(5), 920–940 (2016)
6. Zhu, Y.: Estimating the activity types of transit travelers using smart card transaction data: a case study of Singapore. Transportation, pp. 1–28 (2018)

7. National Bereau of Statics of China, China Statistical Yearbook (2007)
8. Liaw, A., Wiener, M.: Classification and regression by randomForest. R News **2**(3), 18–22 (2002)
9. Cheng, L., et al.: Applying a random forest method approach to model travel mode choice behavior. Travel Behav. Soc. **14**, 1–10 (2019)
10. Breiman, L.: Random forest. Mach. Learn. **45**, 5–32 (2001)
11. Suykens, J.A., Vandewalle, J.: Least squares support vector machine classifiers. Neural Process. Lett. **9**(3), 293–300 (1999)
12. Hastie, T., et al.: Multi-class adaboost. Stat. Interface **2**(3), 349–360 (2009)
13. McFadden, D.: Conditional logit analysis of qualitative choice behavior (1973)

Spatial Algorithms

Maximum Gap Minimization in Polylines

Toni Stankov and Sabine Storandt[✉]

University of Konstanz, Konstanz, Germany
sabine.storandt@uni-konstanz.de

Abstract. Given a polyline consisting of n segments, we study the problem of selecting k of its segments such that the maximum induced gap length without a selected segment is minimized. This optimization problem has applications in the domains of trajectory visualization (see Fig. 1) and facility location. We design several heuristics and exact algorithms for simple polylines, along with algorithm engineering techniques to achieve good practical running times even for large values of n and k. The fastest exact algorithm is based on dynamic programming and exhibits a running time of $\mathcal{O}(nk)$ while using space linear in n. Furthermore, we consider incremental problem variants. For the case where a given set of k segments shall be augmented by a single additional segment, we devise an optimal algorithm which runs in $\mathcal{O}(k + \log n)$ on a suitable polyline representation. If not only a single segment but k' segments shall be added, we can compute the optimal segment set in time $\mathcal{O}(nk')$ by modifying the dynamic programming approach for the original problem. Experiments on large sets of real-world trajectories as well as artificial polylines show the trade-offs between quality and running time of the different approaches.

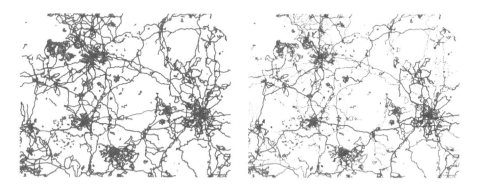

Fig. 1. Cut-out of the real-world trajectory set used for benchmarking. In the left image all segments are drawn. In the right image only $k = \sqrt{n}$ segments are drawn for each trajectory on n segments. This leads to a total reduction from over 19 million segments to be drawn to around 0.8 million segments while still constituting a useful visual representation of the data set.

© Springer Nature Switzerland AG 2020
S. Di Martino et al. (Eds.): W2GIS 2020, LNCS 12473, pp. 181–196, 2020.
https://doi.org/10.1007/978-3-030-60952-8_19

1 Introduction

Visualization of polylines on different levels of granularity is an extensively studied topic, for example in the context of zoom-aware representations of maps. Upon zooming out, the level of detail of the entities displayed in the map is usually reduced in order to avoid visual clutter and to make rendering more efficient. The latter aspect is especially important if data can be dynamically requested to be displayed as a map overlay (e.g. sets of paths or trajectories). The standard way to reduce the complexity of a polyline is simplification. There, a polyline P consisting of n segments is replaced by another polyline P' which consists of fewer segments but still faithfully represents the original polyline according to some distance measure. Polyline simplification comes in many different flavours but all of those approaches share a disadvantage when used for dynamic map rendering: As P' contains segments not present in P, one has to redraw the complete polyline from scratch upon zooming in. Hence accumulated over all zoom levels significantly more than n segments have to be drawn. To avoid this, we suggest an alternative approach for complexity reduction of polylines, namely to only render a selection of the segments in a zoomed-out view and then to fill in the gaps upon zooming in. Formally, a gap is defined as follows.

Definition 1 (Gap). *Given a polyline P and a subset S of its segments, $P \setminus S$ induces a set of subpolylines of P. We call each such subpolyline a gap in P with respect to S.*

For S to be a faithful representation of P, induced gaps should be as small as possible. This leads to the following optimization problem formulation:

Definition 2 (Min Max Gap). *Given a polyline P and a parameter $k \in \mathbb{N}$, select a set S of k segments of P such that the maximum gap length is minimized.*

The min max gap problem also arises in the context of facility location problems where facilities are segments instead of points. For example, a polyline P could model a street where the segment borders indicate crossing points with other streets. Now on k of those segments parking on the side of the road shall be allowed, and the goal is to select them such that the longest drive from any point in P to a selected parking area is minimized.

In this paper, we thoroughly study the min max gap problem from a theoretical and practical perspective.

1.1 Related Work

The idea of displaying only a subset of all polyline segments is conceptually close to many sampling-based approaches which are used to reduce huge data sets in order to make computations and visualizations possible, as investigated e.g. for surfaces [13], volumes [14], or graphs [8]. The insight that gaps in visualizations and drawings might not be a hindrance for correct object perception is also exploited in stipple drawings, see e.g. [9]. Most of these approaches do not come with a clear optimization goal in mind, though, but rather choose the sample points in a (semi-)random fashion.

For polylines, simplification is the conventional way of complexity reduction. One of the most common approaches is to ask for the poyline induced by a subset of the points of the original polyline (including start and end point) that has the fewest possible segments but a (local) distance of at most ε to the original polyline. Typical distance measures are e.g. the Frechet distance or the Hausdorff distance. Frequently used approaches for polyline simplification are the Douglas-Peucker algorithm [3], and the algorithms by Hiroshi and Masao [7] as well as Chan and Chin [1]. The Imai and Iri algorithm runs in $\mathcal{O}(n^3)$; the Chan and Chin algorithm improves this to $\mathcal{O}(n^2)$ for the Hausdorff distance. They are both exact algorithms if the goal is finding the simplified polyline where each segment has distance at most ε to the part of the original polyline it shortcuts. If instead the ε distance applies to the whole original polyline, they turn into heuristics [15]. Then, for the Frechet distance the best known running time of an exact algorithm is $\mathcal{O}(kn^5)$ where k denotes the number of output segments. For the Hausdorff distance the problem even becomes NP-hard [15]. The Douglas-Peucker algorithm is a recursive algorithm which is heuristic in nature. It exhibits a worst-case running time of $\mathcal{O}(n^2)$ but an expected running time of $\mathcal{O}(n \log n)$. When implemented in a more sophisticated way, the worst-case running time can be decreased to $\mathcal{O}(n \log n)$ as shown by Hershberegr and Snoeyink [6]. The polyline simplification variant closer to our problem setting is that given a segment budget, the goal is to find the simplified polyline induced by a subset of the points with the smallest distance to the original polyline. For this variant, a $\mathcal{O}(n^2 \log n)$ algorithm exists [1].

Min max problems are ubiquitous in the realm of facility location problems, including the famous k-center problem [5], where, among a set of n points, k points shall be selected as centers with the goal to minimize the maximum distance of any point to its closest center. There exists a problem variant where the center locations have to be located on a line [16]. In the one-dimensional k-center problem, the input points themselves are located on a line. The problem can then be solved in linear time if the points have no weights (this result even extends to trees) [4], and in $\mathcal{O}(n \log n)$ in case they do [2,11]. But in all 1DkCenter variants, the facilities are always points and not segments. Line-shaped facilities were previously considered in the context of transportation problems [10]. There, a set of n demand points is given and the goal is to identify k line segments in the plane such that the maximum distance of any demand point towards a line segment is minimized (sometimes with an additional constraint on the summed length of all line segments). The difference to our setting is that we cannot choose the line segments freely but have to select them from the given segment set. In [12], an overview of existing work on facility location problems is provided where the demand locations and facilities being each either points, lines or polygons but we are not aware of previous work covering our exact problem formulation.

1.2   Contributions

Given a polyline with n segments and a parameter $k \in \mathbb{N}$, we want to select k of the segments such that the induced gap length is as small as possible. For simple polylines in which all segments have uniform length, we devise a linear time exact algorithm for arbitrary k. For simple polylines with non-uniform segment lengths, we design three exact algorithms: a recursive algorithm, a parametric search algorithm, and a dynamic programming based algorithm. The latter turns out to be the asymptotically fastest with a running time of $\mathcal{O}(nk)$. We also come up with a $(k+1)$-approximation algorithm and a sensible heuristic which run in $\mathcal{O}(n + k \log k)$ and $\mathcal{O}(n)$, respectively. In addition, we consider the incremental version of the problem where already an initial selection of segments is given and the goal is to extend this selection; again with the goal of minimizing the maximum gap length. We provide distinct results for the cases that only a single segment is added or multiple segments at once. For all proposed algorithms, we outline algorithm engineering techniques that are suitable to reduce their running times in practice. We carefully implement the devised exact and heuristic approaches for min max gap and compare them in terms of quality and running time on a large set of polylines, including real-world trajectories as well as artificial benchmarks.

2   Preliminaries and Notation

A polyline P is defined by a list of points or vertices $p_1, p_2, \ldots, p_n$ with $p_i \in \mathbb{R}^2$, and their induced connecting segments $s_i = (p_i, p_{i+1})$ for $i = 1, \ldots, n-1$. The length of a segment is given by $|s_i|$. We use $L(P)$ to refer to the length of P which is the summed length of all contained segments.

Note that the solution for the min max gap problem on simple polylines only depends on the sequence of segment lengths and hence is invariant under transformations which do not affect those. Therefore, we represent the polyline as an array A of length $n-1$ where $A[i] = |s_i|$ for $i = 1, \ldots, n-1$. For ease of exposition, we always assume in the following that the polyline is realized as a straight line. When referring to the subpolyline of P starting at segment s_i and ending at segment s_j, we use $A[i, \ldots, j]$.

3   Simple Polylines with Uniform Segment Lengths

If all segments in P have uniform length, it is unsurprisingly the best to choose the k segments equally distributed over P. To formalize this, we first define the set of k points which induce a subdivision of a continuous realization of P in exactly $k+1$ pieces of equal length.

Definition 3 (Δ_k-points). *Given a polyline P and a parameter k, we set $\Delta_k :=$ $L(P)/(k+1)$ and define the corresponding set of Δ_k-points d_i for $i = 1, \ldots, k$*

as the points on the polyline with distance $i \cdot \Delta_k$ to the starting point p_0. With $S(\Delta_k)$ we refer to the set of segments in P which contain a Δ_k-point.[1]

Lemma 1. *Given a polyline P where all segments have uniform length, and a parameter $k \in \mathbb{N}$, then $S(\Delta_k)$ yields an optimal solution to the min max gap problem.*

Proof. A simple lower bound LB for the min max gap is given by subtracting the summed lengths of the k longest segments in P from $L(P)$ and subsequently dividing by $k + 1$. In case all segments have uniform length l, we get:

$$LB := (L(P) - k \cdot l)/(k+1) = L(P)/(k+1) - k \cdot l/(k+1) > L(P)/(k+1) - l$$

The solution constructed by choosing the segments which contain the Δ_k-points induces a maximum gap of length at most $\Delta_k = L(P)/(k+1)$. The real gap length has to be a multiple of l. As $LB > L(P)/(k+1) - l$ we conclude that $\Delta_k - LB < l$ and hence the bounds match when the LB is rounded up and Δ_k rounded down to a multiple of l; optimality of the solution follows. ∎

Computing the Δ_k-points and selecting the respective segments can be accomplished in time linear in n by performing a left-to-right sweep over the array.

4   Simple Polylines with Arbitrary Segment Lengths

In this section, we propose several heuristic and exact algorithms for the min max gap problem on simple polylines where the segment lengths may differ.

4.1   Uniform Selection Heuristic

It might be tempting to use the selection algorithm proposed above for polylines with uniform segment lengths and apply it also to polylines with non-uniform lengths. This, however, does not even guarantee any finite approximation factor as shown by the example in Fig. 2. The same observation applies to the strategy of simply selecting every $\lceil n/k \rceil$th segment.

4.2   Greedy Approximation Algorithm

If our optimization goal would be to select k segments such that the sum of all gap lengths is minimized, then the optimal solution would be to select the k longest segments in P. For our min max gap problem this might be a bad choice, though, as those segments might form one block at the beginning or end of the polyline and hence induce a long gap. However, in contrast to the uniform selection heuristic, this gap can not be arbitrarily longer than the optimum gap as shown in the following lemma.

---

[1] In case a Δ_k-point coincides with a segment border, one of the two possible segments is chosen arbitrarily into the set $S(\Delta_k)$.

Fig. 2. Given the polyline $P = p_1, p_2, p_3, p_4$ and $k = 2$, the optimal solution would be to select the segments (p_1, p_2) and (p_3, p_4) resulting in a gap length of ε. The uniform selection strategy however could choose the segments (p_1, p_2) and (p_2, p_3) as they contain the Δ_2-points d_1 and d_2. The resulting gap length would be $\Delta_2 = L(P)/3$ and hence for $\varepsilon \to 0$ arbitrarily worse compared to the optimal min max gap.

Lemma 2. *Greedy selection of the k longest segments in P leads to a min max gap length of at most $k + 1$ times the optimum.*

Proof. Let L_k be the accumulated length of the k longest segments in P. As argued in the proof of Lemma 1, $(L(P) - L_k)/(k + 1)$ is a valid lower bound for the min max gap. The gap induced by the greedy algorithm has a length of at most $L(P) - L_k$; the lemma follows. ∎

The running time of this greedy approach is $\mathcal{O}(n + k \log k)$.

4.3 Naive Exact Selection Algorithm

There is a straightforward exact algorithm for the min max gap problem: Testing all segment subsets of size k and keeping track of the min max gap. There are $\binom{n-1}{k}$ such sets. The max gap length computation requires linear time for each set, hence we end up with a total running time of $\mathcal{O}\left(n^{1+\min\{k, n-k\}}\right)$ which is clearly prohibitive for practical use.

4.4 Parametric Search

Parametric search was applied to related problems before [2,4], as it neatly allows to transform an answer to the decision version of a problem into an answer to the optimization version of the problem. Here, the decision version asks for a given gap length G whether it can be realized by choosing $\leq k$ segments from P. If an oracle answering that question is available, the min max gap problem can be solved by systematically testing all possible gap lengths. For designing such an oracle, we exploit the following definition and the subsequent lemma:

Definition 4. (Maximal Gap). *For a given gap length G, we call a gap between segments s_i and $s_{j>i}$ in P maximal if $L(A[i + 1, \ldots, j - 1]) \leq G$ but*

$$L(A[i, \ldots, j - 1]) > G$$

and

$$L(A[i + 1, \ldots, j]) > G.$$

Lemma 3. *Let G^* be the min max gap for given P and k, then there exists S^* with $|S^*| \leq k$ which realizes G^* and the first k gaps induced by S^* are maximal.*

Proof. Let S_0 be any segment set of size at most k which realizes G^*. Let $s_i \in S_0$ be the first segment (in the order of appearance along P) that is not the right border of a maximal gap. Then we could construct $S_1 = s_{i+1} \cup S_0 \setminus s_i$ and we would still have a valid segment selection that realizes G^*. Iteratively applying this argument then allows to construct a solution in which all segments are right borders of a maximal gap. Note that in case we have at some point $s_i = s_{n-1}$ this could also result in a reduction of the segment set. ■

According to the lemma, for a given max gap length G, the following sweep algorithm identifies a segment set S^* with $|S^*| \leq k$ which realizes that gap: Iterate over the array A representing P from left to right. As soon as the summed segment lengths exceed G, the current segment is included in S^* and the process is repeated starting from the next segment. This algorithm runs in time $\mathcal{O}(n)$. To find the optimal G, we need to test all possible gap lengths, that is, the lengths of all $\mathcal{O}(n^2)$ subpolylines of P. This results in a total running time of $\mathcal{O}(n^3)$. If the subpolyline lengths are sorted decreasingly, we can stop as soon as the size of the produced segment set S^* becomes larger than k; then the gap length investigated prior to the current one constitutes the optimal min max gap. Alternatively, a binary search over the sorted potential gap lengths produces the optimal solution in time $\mathcal{O}(n^2 \log n)$.

For faster practical implementation we can use the observation that the min max gap length can be no longer than Δ_k (as this gap length is realized by the uniform selection heuristic). Hence we only need to check subpolyline lengths smaller than or equal to Δ_k.

In case the segment lengths are all natural numbers, at most $L(P)$ gap lengths need to be checked. Using binary search again, this results in a running time of $\mathcal{O}(n \cdot \log L(P))$ for solving the min max gap problem exactly.

4.5   Recursive Selection Algorithm

As shown above, the uniform selection algorithm does not guarantee to find an optimal solution because not all segments in the optimal solution have to contain a Δ_k-point. The following lemma shows, though, that there still is an interesting connection between the optimal solution and the Δ_k-points.

Lemma 4. *Let P be a polyline and S^* a set of segments that realize the optimal min max gap for given parameter $k \in \mathbb{N}$. Then at least one segment in S^* contains a Δ_k-point.*

Proof. The uniform selection heuristic leads to a min max gap of length at most $\Delta_k = L(P)/(k+1)$. In case the uniform selection yields an optimal min max gap, the lemma is trivially true. Lets assume otherwise. Then the optimal min max gap has to be strictly smaller than Δ_k. Therefore, if we look at the $k + 1$

Algorithm 1: RecMinMaxGap(P,k)

Input: polyline P, parameter $k \in \mathbb{N}$
Output: optimal min max gap G
1 $S \leftarrow$ set of segments in P containing a Δ_k-point;
2 **if** $k = 0$ **then**
3 | **return** $L(P)$;
4 **end**
5 **if** $k \geq |P|$ **then**
6 | **return** 0;
7 **end**
8 $G \leftarrow L(P)$;
9 **for** $s = (p_i, p_{i+1}) \in S$ **do**
10 | $P_1 = p_1, \ldots, p_i$;
11 | $P_2 = p_{i+1}, \ldots, n$;
12 | **for** $j=0$ to k-1 **do**
13 | | $l \leftarrow$ RecMinMaxGap(P_1,j);
14 | | $r \leftarrow$ RecMinMaxGap(P_2,k-1-j);
15 | | $g \leftarrow \max(l, r)$;
16 | | $G \leftarrow \min(G, g)$;
17 | **end**
18 **end**
19 **return** G;

---

subpolylines of length Δ_k induced by the Δ_k-points, we know that the optimal solution needs to contain at least one segment intersecting each subpolyline. But as the optimal solution is only allowed to contain k segments and needs to cover $k+1$ subpolylines, it follows that at least one segment needs to intersect at least two of the subpolylines at once. This segment then automatically contains the Δ_k-point at their intersection. ∎

Hence we know that among the set of at most k segments in P which contain the Δ_k-points, at least one has to be contained in an optimal solution set. This insight is crucial to develop an algorithm which first guesses which of the segments it is and then recurses on the two subpolylines left and right of the selected segment. The detailed pseudo-code is given in Algorithm 1. Note that the algorithm returns the min max gap length G^* instead of corresponding segment set S^*. But as argued in the context of parametric search, once we know G^* we can compute S^* in linear time.

Lemma 5. *The recursive min max gap algorithm runs in* $\mathcal{O}(2^k(k!)^2 n)$.

Proof. The recursive running time for an input polyline on n points and for fixed k can be upper bounded as follows:

$$T(n,k) \leq k \cdot \sum_{j=0}^{k-1}(T(n-1,j)+T(n-1,k-1-j))+cn \leq 2k^2 \cdot T(n-1,k-1)+cn$$

Expanding the formula k times completes the proof. ∎

In contrast to the naive exact algorithm, the running time of the recursive algorithm is exponential in k but linear in n. Hence for any $k \in \mathcal{O}(1)$ we get a linear running time. Note that without the insight provided in Lemma 4, we would need to recurse on all segments in P and hence get a running time exponential in n. Nevertheless, the exponential growth in k is still prohibitive for practical use in case k is not a small constant.

To boost practical performance, we observe that any segment with an individual length $>\Delta_k$ always has to be included in S^* (as otherwise the induced gap would be larger than Δ_k and hence worse than the solution of the uniform selection heuristic). This of course can also be applied to deeper levels in the recursion. Hence whenever we find such a segment for a given subpolyline and given k, we don't need to recurse on all possible choices of Δ_k-points containing segments but can just use this particular segment right away to split the subpolyline further.

Also, if for some choice of j in line 12 of the recursive algorithm the left gap l is smaller than the right gap r, then obviously allowing a larger segment budget on the left and a lower segment budget on the right won't lead to a smaller min max gap. Hence in that case no larger values of j need to be investigated.

4.6   Dynamic Programming (DP)

The recursive algorithm is wasteful as it recomputes partial solutions for the same subpolylines and the same value of k over and over again. To avoid those recomputations, we can store those partial solutions in a look-up table and hence transform the recursive algorithm into dynamic programming algorithm. The dynamic programming table has $k+1$ rows, numbered from 0 to k, and $\mathcal{O}(n^2)$ columns, where each column refers to a contiguous subpolyline of P, sorted increasingly by number of contained segments. The table gets filled from top to bottom, left to right, putting in cell (i, j) an optimal set S^* with $\leq j$ segments from the subpolyline in column i such that its max gap is minimized. The induced gap length G is stored along. The time necessary to compute the entry of a cell amounts to $\mathcal{O}(n + k^2)$, leading to a total running time of $\mathcal{O}(k^3 n^2 + k n^3)$ – which unfortunately is larger than the running time of the parametric search approach.

But inspired by Lemma 3 (which we used originally to construct an oracle for parametric search), we can design a more efficient dynamic program: We use $k+1$ rows as before, but now only n columns. In cell (i, j), we put the min max gap $G(i, j)$ for the prefix of P up to segment i which is realized by a segment set of size at most j which contains s_i. For all cells with $j \geq i$ the entry is 0. For $j = 0$, we get $G(i, 0) = L(A[1, \ldots, i])$. The other cells can be filled from top to bottom, left to right, using the following formula:

$$G(i, j) := \min_{l=1,\ldots,i-1} \max\{G(l, j-1), L(A[l+1, \ldots, i-1])\}$$

The final min max gap is then computed by applying the formula to $G(n, k+1)$.

Lemma 6. *The DP algorithm computes the optimal min max gap.*

Proof. We prove the lemma by induction over k. For $k = 1$, we get $G(i, 1) = \sum_{l=1}^{i-1} |s_l|$ which obviously is the smallest possible gap left of segment s_i if only segment s_i is part of the solution. Now lets assume that for all cells $(i, \leq k)$ the optimal gap length to the left $G(i, \leq k)$ is known and we next consider

$G(i, k + 1)$. Let S^* be the optimal solution to the min max gap problem on $A[1, \ldots, i - 1]$ for parameter k that obeys the characteristics of Lemma 3, and let $s_l \in S^*$ be the segment with highest index in that solution set. Then we know that selecting other segments $S^* \setminus s_l$ can not decrease the gap length left of s_l by definition of maximality of the induced gaps. Hence we know that $G(l, k)$ equals this optimal min max gap length left of s_l. For $G(i, k + 1)$ it then follows that with $\max\{G(l, k), L(A[l + 1, \ldots, i - 1])\}$ we consider the min max gap induced by the optimal solution S^*. Correctness follows. ∎

The space consumption of the table is $\mathcal{O}(nk)$, naively. But we observe that the formula to fill a cell only depends on the cells in the previous row. Hence it is sufficient to keep row $j - 1$ in memory to fill out row j, reducing the space consumption to $\mathcal{O}(n)$. Note that we also do not require the whole table to backtrack the optimal segment set as we can again simply use the sweep algorithm discussed for parametric search to get S^* once G^* is known.

The naive running time bound would be $\mathcal{O}(n^2 k)$ as in the worst case we need to look at all entries in row $j - 1$ to fill the last cell in row j. Hence filling a single cell can cost up to $\mathcal{O}(n)$. However, we make use of the following lemma to deduce a better bound:

Lemma 7. *If the value in $G(i, j)$ equals $\max\{G(l, j - 1), L(A[l + 1, \ldots, i - 1])\}$, then the value of $G(i + 1, j)$ can be determined by $\max\{G(l', j - 1), L(A[l' + 1, \ldots, i])\}$ for some $l' \geq l$.*

Proof. Assume $l' < l$ for contradiction. It follows that $G(l, j - 1) > L(A[l + 1, \ldots, i])$ as otherwise $G(i + 1, j)$ would assume the value $L(A[l + 1, \ldots, i])$ for l and we would get $L(A[l' + 1, \ldots, i]) > L(A[l + 1, \ldots, i]) > G(i + 1, j)$ for all $l' < l$. But if $G(l, j - 1) > L(A[l + 1, \ldots, i])$ we also know $G(l, j - 1) > L(A[l + 1, \ldots, i - 1])$ and hence $G(i, j) = G(l, j - 1)$. As $G(i + 1, j) = G(l, j - 1) = G(i, j)$ and $G(i + 1, j) \geq G(i, j)$ we know that there cannot exist $l' < l$ such that $G(i + 1, j) = \max\{G(l', j - 1), L(A[l' + 1, \ldots, i])\} < G(l, j - 1)$ which concludes the proof. ∎

In addition, we observe that if $G(l, j - 1)$ is the value that determines $G(i, j)$, we only need to investigate values up to $G(l + 1, j - 1)$. This is due to $G(l, j - 1)$ monotonically increasing for growing l but $L(A[l + 1, \ldots, i])$ decreasing at the same time. Therefore, the smallest max value of those two is either directly before or directly after the point where $L(A[l + 1, \ldots, i])$ becomes larger than $G(l, j - 1)$. Using this observation together with the lemma, we can conclude that we need time $\mathcal{O}(n)$ to fill the whole row rather than just a single cell. The resulting overall running time of this DP approach is hence $\mathcal{O}(nk)$.

We summarize these findings in the following theorem.

Theorem 1. *The min max gap problem can be solved in time $\mathcal{O}(nk)$ with a space consumption of $\mathcal{O}(n)$.*

5   Incremental Segment Selection

In our envisioned application of zoom-aware rendering of trajectories, we first solve the min max gap problem for the most zoomed-out view. Then, upon zooming in, we want to augment the set of rendered segments instead of rendering a new solution to the min max gap problem with increased segment budget from scratch. This implies that we need to be able to solve incremental variants of the min max gap problem.

5.1   Incremental Addition of Single Segments

We first consider the case where only one additional segment needs to be selected. Lets assume that we already have a selection of $k - 1$ segments and now want augment this selection by choosing a kth segment.

Lemma 8. *Given a polyline P and a subset S of its segments of size $k - 1$; the optimal min max gap induced by any superset S^+ of S of size k is realized by identifying the currently largest gap G in P with respect to S and then choosing the segment which contains the Δ_1-point of G.*

Proof. Let G^+ be the best min max gap achievable for a set $S^+ \supset S$ with $|S^+| = k$. It is obvious that a segment intersecting G needs to be selected to realize $L(G^+) < L(G)$. If G^+ is a subpolyline of G, then not selecting a segment which contains the Δ_1 point of G induces $L(G^+) > L(G)/2$ while the choice of the middle segment ensures $L(G^+) \leq L(G)/2$. Hence the latter is always preferrable. In case G^+ is not a subpolyline of G, we always have $L(G^+) \geq L(G)/2$. Hence again, choosing a segment in G which contains the Δ_1-point allows to realize the min max gap G^+. ∎

The running time for one incrementation is hence linear in n. However, if we want to repeatedly add single segments, we can actually do better by changing the representation of the polyline: Instead of using the array A with $A[i] = |s_i|$ we now use the array B with $B[i] = \sum_{j=1}^{i} |s_i|$. Given A, we can compute B in linear time. Once B is available, we can compute the gap length between any two segments in constant time. Hence the overall longest gap can be identified in time $\mathcal{O}(k)$ for the $k - 1$ given segments (if they are already sorted increasingly by their index). The search for the middle segment within this gap then boils down to a binary search on the respective part of B and hence takes time $\mathcal{O}(\log n)$. Therefore, an incrementation of the segment set by a single additional segment requires only time $\mathcal{O}(k + \log n)$ on representation B.

Note that we can also utilize this result to solve our original min max gap problem by selecting all k segments incrementally; starting with the optimal solution for one segment and then always augmenting it until k segments are selected. This leads to an overall running time of $\mathcal{O}(nk)$ when using representation A and $\mathcal{O}(k \log n + k^2)$ on representation B (plus a linear time preprocessing to compute B). Unfortunately, this again is only a heuristic for the min max gap problem.

5.2   Incremental Addition of Segment Sets

If we want to select $k' > 1$ segments at once to augment an existing set S of size k, we have to decide how many segments to spend on each of the at most $k + 1$ currently existing gaps. Naively we can test all possible distributions of the k' segments over the gaps, compute for each gap and the respective segment budget the optimal solution using one of the exact algorithms described above, and then keep track of the best solution. Computing for all gaps and all budgets $\leq k'$ the optimal solution requires time $\mathcal{O}((n - k)k')$ in total when using the DP approach. But checking all possible partitions of k' into $k + 1$ parts takes an impractical time of $\Theta(p_{k+1}(k'))$.

But we can do better by modifying the DP approach as follows: For each segment s_i we store a pointer to $s_l \in S$ with l being the largest index $< i$ (we point to a dummy segment s_0 of length 0 in case such an l does not exist). When we fill in a cell (i, j) with $s_i \notin S$, we have to distinguish two cases: either the best solution is to select all segments other than s_i left of s_l or at least one of those segments right of s_l. In the first case the resulting gap is $\max\{G(l, j - 1), L(A[l + 1, \ldots, i - 1])\}$. In the second case, we can apply the original formula to the cells $(l', j - 1)$ for $l' = l + 1, \ldots, i - 1$. Checking which of the two cases leads to the smaller max gap suffices to compute $G(i, j)$.

When we fill in a cell (i, j) with $s_i \in S$, we know for sure that all j segments have to be selected left of s_i as s_i can not be selected again. Then, like above, we distinguish whether all those segments are selected left of s_l or at least one of them right of s_l. In the first case the resulting gap is $\max\{G(l, j), L(A[l + 1, \ldots, i - 1])\}$ (note that we now access values in the same row instead of row $j - 1$). In the second case, we apply the original formula but now to the cells (l', j) with $l' = l + 1, \ldots, i - 1$ to get the correct result.

The observations that we used in the analysis of the original DP approach to show that a whole row can be filled in time $\mathcal{O}(n)$ still apply here. Hence we can solve this incremental problem version to optimality in time $\mathcal{O}(nk')$, that is, independent of k.

6   Experimental Evaluation

We implemented all proposed algorithms in C++. Experiments were conducted on a single core of an Intel i5-4300U CPU with 1.90 GHz and 12 GB RAM.

6.1   Benchmarks

We extracted 20.000 trajectories from the set of publicly available GPS traces on OpenStreetMap[2] with different segment numbers and sampling densities. The total number of segments in the set is roughly 19 million. We used a length precision of 1 dm. Figure 3 shows a scatter plot of minimum/maximum segment length of all trajectories on the left, and the distribution of segment numbers on the right. The average n is 953.

---

[2] https://www.openstreetmap.org/traces.

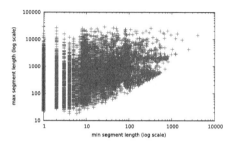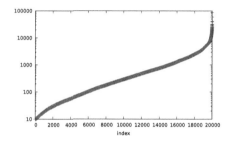

Fig. 3. Max/min segment lengths (left) and number of segments (right) for each trajectory.

In addition, we constructed artificial inputs by creating arrays A in a random fashion. In particular, segments lengths were chosen with uniform probability from the set $\{1, \ldots, 1.000\}$. For each value of n considered in the experiments, we produced a set of 10.000 such polylines.

6.2   Exact Algorithms

We first compared the running times of the three exact approaches on an artificial benchmark with $n = 1.000$, see Fig. 4. As expected, the running time of the recursive approach skyrockets quickly with growing k; results for $k > 10$ could not be produced. But the parametric search and the DP approach perform well (query times of at most 25 ms) for all choices of k. Somewhat surprisingly, the parametric search time even declines with growing k. This can be explained by our use of Δ_k as an upper bound for the optimal min max gap length. As Δ_k shrinks with growing k the set of gap lengths that need to be considered in the binary search procedure shrinks as well. The running time of the DP approach increases with growing k, as there need to be more rows filled in the DP table. Therefore, parametric search becomes faster than DP for large k. The

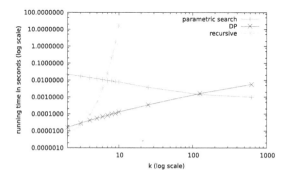

k	PS	DP
10	0.807	0.001
50	0.242	0.006
250	0.119	0.036
1250	0.096	0.158
6250	0.091	0.661

Fig. 4. Running times of the exact approaches on polylines with $n = 1.000$.

Fig. 5. Running times (in seconds) of parametric search (PS) and DP for $n = 10.000$.

same behaviour can be observed for larger n, as shown in Fig. 5. Using DP up to $k \approx 0.1 \cdot n$ and switching to parametric search for higher k results in query times of less than $100\,\text{ms}$ for all k for $n = 10.000$.

On the real-world benchmark, we chose k for each trajectory individually. We tested two settings: $k_1 = \sqrt{n}$ using DP and $k_2 = n/2$ using parametric search. For k_1, DP took on average less than $1\,\text{ms}$ per trajectory (maximum $310\,\text{ms}$), and for k_2, parametric search took on average $5\,\text{ms}$ per trajectory (maximum $7239\,\text{ms}$). Hence although the average performance is satisfying it might be sensible to switch to heuristics for trajectories with a huge number of segments to decrease the peak times.

6.3   Heuristic Algorithms

Next, we evaluated the running time and the quality of the proposed heuristics, namely the uniform selection heuristic, the greedy approximation algorithm and the incremental heuristic.

On the real-world benchmark, we ran the three approaches for $k_1 = \sqrt{n}$ and $k_2 = n/2$ as above but in addition also for $k_0 = 10$ and $k_3 = 4n/5$ to get a more complete picture, see Table 1 for the quality results. We observe that the uniform selection heuristic performs best in terms of quality but also in terms of running time. The maximum time to process a trajectory was well below $1\,\text{ms}$ and hence three to five orders of magnitude faster than the exact approaches. Nevertheless, the quality deteriorates for very small or large values of k. For small k, the chance to unluckily only select very short segments is higher. For large k, it gets more likely that longer segments contain more than one Δ_k-point which results in less than k segments being selected in total. The incremental heuristic is the slowest (with an average of $24\,\text{ms}$ per trajectory for k_3) and exhibits for large k also the worst quality. But for its original purpose (augmenting a given segment set by one additional segment) it is optimal and takes only about a microsecond. The greedy approximation algorithm runs in less than a millisecond on each trajectory. But, as expected, it suffers from too long gap lengths on average and in the worst case. Only for small values of k its approximation guarantee is helpful: For k_0 we see that the guarantee of a factor of 11 is almost maxed out

Table 1. Quality measured as computed max gap length divided by optimal max gap length for the three heuristics on the set of real-world trajectories.

k	Uniform			Incremental			Greedy		
	Min	Avg	Max	Min	Avg	Max	Min	Avg	Max
10	1.00	1.51	665.46	1.02	4.28	3066.71	1.00	4.17	10.95
$\sqrt{n}$	1.00	1.05	6.25	1.00	1.45	40.51	1.00	7.37	186.98
$0.5n$	1.00	1.65	156.51	1.00	8.35	6066.33	1.00	20.54	14637.50
$0.8n$	1.00	4.23	2402.99	1.04	41.27	19379.25	1.00	10.88	3184.36

on the benchmark but it is still results in a better factor compared to the other two approaches.

7 Conclusions and Future Work

We discussed several approaches to tackle the min max gap problem on simple polylines. As (optimal) solutions can be computed quickly, we can reduce the complexity of trajectories on-the-fly to allow for faster visualisation (or to speed-up other kinds of computations on the trajectory). For trajectory sets, there are several advantages of drawing a subset of the segments rather than simplified trajectory versions with shortcuts: As outlined in the introduction, we draw less segments in total aggregated over all zoom levels. But we also avoid visual artefacts which could arise when trajectories partially overlap but their simplifications do not. Furthermore, we avoid topological inconsistencies as e.g. a pedestrian trajectory cutting across a body of water due to simplification or a city now being on the other side of a trajectory because of a shortcut. In future work, it might be interesting to consider the whole set of trajectories at once and not every trajectory individually to avoid drawing similar segments multiple times while having long gaps elsewhere; and to extend the results to polygons and non-simple polylines in general. Furthermore, other quality measures than the induced max gap length could be used for sampling as e.g. the Hausdorff distance between the complete polyline and the selected segments. Finally, also combinations of simplification and sampling could be useful for practical applications.

References

1. Chan, W.S., Chin, F.: Approximation of polygonal curves with minimum number of line segments or minimum error. Int. J. Comput. Geom. Appl. **6**(01), 59–77 (1996)
2. Cole, R.: Slowing down sorting networks to obtain faster sorting algorithm. In: 25th Annual Symposium on Foundations of Computer Science 1984, pp. 255–260. IEEE (1984)
3. Douglas, D.H., Peucker, T.K.: Algorithms for the reduction of the number of points required to represent a digitized line or its caricature. Cartogr. Int. J. Geogr. Inf. Geovisualization **10**(2), 112–122 (1973)
4. Frederickson, G.N.: Parametric search and locating supply centers in trees. In: Dehne, F., Sack, J.-R., Santoro, N. (eds.) WADS 1991. LNCS, vol. 519, pp. 299–319. Springer, Heidelberg (1991). https://doi.org/10.1007/BFb0028271
5. Gonzalez, T.F.: Clustering to minimize the maximum intercluster distance. Theor. Comput. Sci. **38**, 293–306 (1985)
6. Hershberger, J., Snoeyink, J.: An $O(n\log n)$ implementation of the Douglas-Peucker algorithm for line simplification. In: Proceedings of the Tenth Annual Symposium on Computational Geometry, pp. 383–384. ACM (1994)
7. Hiroshi, I., Masao, I.: Polygonal approximations of a curve-formulations and algorithms. Mach. Intell. Pattern Recogn. **6**, 71–86 (1988)

8. Leskovec, J., Faloutsos, C.: Sampling from large graphs. In: Proceedings of the 12th ACM SIGKDD International Conference on Knowledge Discovery and Data Mining, pp. 631–636. ACM (2006)

9. Lu, A., Morris, C.J., Ebert, D.S., Rheingans, P., Hansen, C.: Non-photorealistic volume rendering using stippling techniques. In: Proceedings of the Conference on Visualization 2002, pp. 211–218. IEEE Computer Society (2002)

10. MacKinnon, R.D., Barber, G.M.: A new approach to network generation and map representation: the linear case of the location-allocation problem. Geogr. Anal. $4(2)$, 156–168 (1972)

11. Megiddo, N., Tamir, A., Zemel, E., Chandrasekaran, R.: An $O(n\log^2 n)$ algorithm for the kth longest path in a tree with applications to location problems. SIAM J. Comput. $10(2)$, 328–337 (1981)

12. Miller, H.J.: GIS and geometric representation in facility location problems. Int. J. Geogr. Inf. Syst. $10(7)$, 791–816 (1996)

13. Tanaka, S., Shibata, A., Yamamoto, H., Kotsuru, H.: Generalized stochastic sampling method for visualization and investigation of implicit surfaces. Comput. Graph. Forum 20, 359–367 (2001)

14. Theußl, T., Moller, T., Groller, M.E.: Optimal regular volume sampling. In: Proceedings Visualization 2001. VIS 2001, pp. 91–546. IEEE (2001)

15. van Kreveld, M., Löffler, M., Wiratma, L.: On optimal polyline simplification using the Hausdorff and Fréchet distance. arXiv preprint arXiv:1803.03550 (2018)

16. Wang, H., Zhang, J.: Line-constrained k-median, k-means, and k-center problems in the plane. Int. J. Comput. Geom. Appl. $26(03n04)$, 185–210 (2016)

A Non-cooperative Game Approach for the Dynamic Modeling of a Sailing Match Race

Lamia Belaouer[1], Mathieu Boussard[2], Patrick Bot[3], and Christophe Claramunt[3(✉)]

[1] E-Cobot, Nantes, France
[2] Craft Ai, Paris, France
[3] Naval Academy Research Institute, Lanvéoc-Poulmic, France
christophe.claramunt@gmail.com

Abstract. This research introduces a spatio-temporal planning framework whose objective is to simulate a sailing yacht match race. The race is a duel in which strategy and tactics play a major role as sailors continuously have to take decisions according to wind variations and opponent's locations and actions. We introduce a decision-aid framework based on a stochastic game approach grounded on an action-oriented model that replicates yachts' behaviors. The objective is to replicate as closely as possible the respective behaviors and navigation decisions taken by yachts competitors. The proposed formalism has been implemented and is illustrated by a sample race example.

Keywords: Sailing match race · Game theory · Dynamic routing

1 Introduction

The research presented in this paper models the dynamics of a sailing race between two yachts with similar performances. These yachts compete against each other with racing rules, and where tactical decisions from the skippers as well as the physical environment (i.e., wind) are the constraints and actions considered, those being partly expected and partly unexpected. In order to win a race, not only sailors should be sufficiently skilled, but also they should take the most appropriate decisions at the right times taking into account the opponent's behavior and wind fluctuations during the race. This clearly emphasizes the predominance of the strategy dimension [1]. The best strategy is surely not to optimize the route as applied in conventional weather routing, but rather to maximize the probability to reach the finish line before the opponent and by taking into account the opponent's behavior [6]. Wind conditions vary over time, while actions are tacking decisions taken by sailors during the course. A tacking is a sailing maneuver so that the direction from which the wind blows changes from one side of the yacht to the other.

© Springer Nature Switzerland AG 2020
S. Di Martino et al. (Eds.): W2GIS 2020, LNCS 12473, pp. 197–213, 2020.
https://doi.org/10.1007/978-3-030-60952-8_20

In fact a yacht race can be considered as a decision-making process that generalizes a *Markov Decision Process* (MDP) [7].In a series of related work, wind variations have been modeled as a MDP process [2–6]. So far these models have not been applied to yacht races where the opponent behavior should be also taken into account [6]. Our objective is to take into account variations of wind directions to compute the most appropriate tacking and heading directions at appropriate locations so as to minimize expected arrival time. The second component of our approach retains the principles behind game theory [11]. We model a yacht race as a stochastic game in which probabilistic actions played by the players as well as wind fluctuations are taken into account [9]. Overall the game provides a sequence of states. At each state every player (i.e., the two yachts) selects some actions and receives a payoff that depends on the current state and the chosen actions, and so on until the end of the race. The remainder of paper is organized as follows. Section 2 presents the main principles of the sailing match race. Section 3 develops the modeling approach. Section 4 introduces some experimental evaluations. Section 5 briefly describes related work. Finally, Sect. 6 concludes the paper and outlines further work.

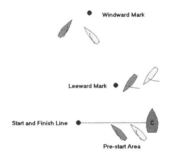

Fig. 1. Schematic representation of a yacht race

2   Sailing Race Principles

Nowadays, yacht match racing is a popular form of sailing competition where two boats compete against each other. A race is made of several legs. Every leg is delimited by a buoy at the start and the end. Every buoy has to be rounded several times so most often the start and finish lines are identical. Therefore, the yachts sail upwind and downwind, alternatively (Fig. 1). The respective distances to the windward marks from the start or end lines are measured in time units. Race committees set the distance to the windward mark so that two laps will be completed in approximately 25 to 35 min. A course is influenced by wind and wave conditions and the types of yachts being sailed. For instance, a crew can force their opponents towards a disadvantageous position by means of tactical craftsmanship. Overall, the winner is generally faster and very likely to make more maneuvers than a fleet race that will sail in similar conditions. The objective of weather routing is to find the fastest route taking into account the yacht performance and wind conditions (i.e., direction and strength). In an upwind leg, a yacht cannot sail straight into the wind and has to zig-zag towards the upwind mark. In order to reach the upwind mark (windward mark), the yacht's skipper must perform a maneuver known as tacking which turns the bow of the yacht through the wind. This maneuver slows down the yacht, so the number of tacks should be minimized. In a constant and homogeneous wind, the optimal route includes only one tack. However, most often the wind direction and speed vary in space and time, and the optimal route includes several tacks to take advantage of the wind shifts, provided the gain exceeds the loss due to tacking. Let us introduce two typical sample configurations (Fig. 2):

- The wind is continuously shifting to one side -say right- during the upwind leg; the optimal route includes one tack and starts going right, as the opposite strategy -going left- leads to a longer path (Fig. 2(a)).
- The wind oscillates with a shift to the right, then shifting back to its initial direction, then shifting to the left and then shifting back to its initial direction, during the upwind leg The optimal route includes one tack and starts going left while the wind is shifted to the right, then tacking when the wind is shifted to the left, to maximize the projected velocity towards the upwind mark (Fig. 2(b)). The yacht should not tack as soon as the wind starts to shift back from the maximum right shift, but when the wind crosses the initial direction and becomes shifted to the left of the mean wind direction.

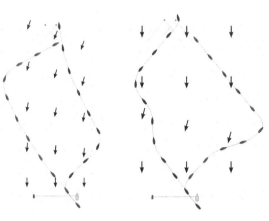

(a) Wind continuously shifts to the right (veering) during the upwind leg.

(b) Wind oscillates once back and forth during the upwind leg, shifting to the right and then back to its initial direction.

Fig. 2. Weather routing and wind fluctuations in two basic cases

These sample cases illustrate how a winning strategy depends on the expected wind and relies upon foreseeing the wind shifts. During a continuous wind shift, one should first sail on the unfavorable tack, while during a wind oscillation, one should always sail on the favorable tack. The winning strategy may include several tacks to take advantage of the wind shifts, provided the gain exceeds the loss due to tacking. Moreover, skippers may take advantage of wind speed variations. Yacht performance generally increases with wind speed (except in extreme circumstances), so the best strategy also includes going into stronger wind areas and avoiding wind holes. Wind speed variations may be located in time and space —gusts or holes moving on the race course– or stationary wind heterogeneity may exist on the race course due to land or temperature gradients. A naive strategy is to search for the fastest route to reach the upwind mark. However, when racing against an opponent in an uncertain wind, the opponent's behavior should be considered and the best strategy should take into account the opponent's behavior as well as the wind fluctuations in order to reach the finish line before the opponent [6]. Yachts relative locations and behaviors are key elements during the race. Yachts have to avoid collision and to obey racing rules given by the International Sailing Association. For instance, when boats are on opposite tacks (tack: side the yacht is receiving the wind from), a port-tack boat shall keep clear of a starboard-tack boat (Fig. 3(a)) (the port, resp. the starboard, denotes the left, resp. right, side of a boat when looking forward towards the bow). When boats are on the same tack and overlapped, a windward boat shall keep clear of a leeward boat (Fig. 3(b)) (to tack: to turn the boat across the wind from one tack to the other; leeward: the side the wind is blowing to; windward: the side the wind is blowing from).

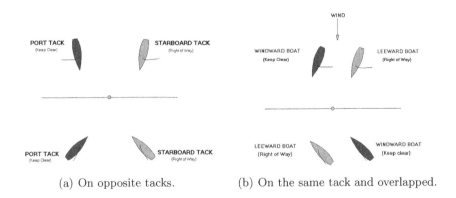

(a) On opposite tacks.   (b) On the same tack and overlapped.

Fig. 3. Collision avoidance example

These sample cases illustrate how the strategy of one sailor depends on the opponent's behavior. Loose covering occurs when the leading yacht sails so as to minimize the lateral distance to the opponent, or to keep a position between the opponent and the upwind mark in order to minimize the risk of the opponent making a gain through a subsequent wind shift. When yachts are close to each other, the presence of the opponent also alters each yacht performance as the wind field is perturbed by a sailing yacht [14, 15].

3 Modelling Approach

Without loss of generality, let us consider an upwind leg (Fig. 4). The laylines are the borders of the race course. The closer a yacht get to the laylines (layline: the line to sail straight to the mark, without tacking, at the optimal VMG), the less flexibility it has to take advantage of any wind fluctuation (even a yacht might try to push the opponent outside the layline). Overall, the winner is the one who reaches the upwind mark first. The modeling approach developed considers two main components: the weather- and the action-based dimensions.

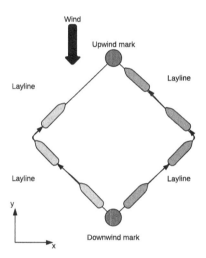

Fig. 4. Schematic representation of the course. The y axis is oriented in the direction of the upwind mark (initial wind direction)

The wind model considered is the one developed in [6] that combines weather forecast interpolated on one-minute time step, a short-term forecast based on Artificial Neural Networks fed with past observations on the yacht, and a stochastic model using a Markov process. Yacht performance and crew ability are considered as equivalent. Yacht performance and the crew ability to trim and drive the yacht are not taken into account. We consider identical yachts always sailing upwind at their best VMG (Velocity Made Good, i.e., maximal yacht velocity projected towards the wind direction [6]) so the yacht speed and wind angle are functions of wind speed and the type of yachts considered and are determined by the polar performance diagram which is assumed to be known. Local wind perturbations as well as collision-avoidance actions are not considered at this stage as the race modeled considers distant yachts at the start of the race in order to minimize risk collisions. Each yacht has two possible courses —on opposite *tacks*— at a given speed and the only possible actions are to tack or not to tack. A sailing match racing is hereafter modeled as a stochastic game as initially introduced by Shapley [9], and that proceeds step by step from the respective locations of the yachts and according to transition probabilities that reflect successive wind fluctuations and the yachts' actions taken. More formally, a stochastic game is a tuple $\langle Ag, \{A_i : 1, \ldots, n\}, \{R_i : 1, \ldots, n\}, S, T\rangle$, where:

- Ag is a finite set of players indexed $1, \ldots, n$;
- A_i is a finite set of actions available to player i and $\langle a_1, \ldots, a_n\rangle \in \times_i A_i$ denotes a vector of the joint action of the players;
- $R_i : S \times \overrightarrow{A} \to \{-1, 0, 1\}$ is a reward function for the ith player;
- S is a finite set of states;
- $T : S \times \overrightarrow{A} \times S \to [0, 1]$ is a transition function. It indicates the probability of moving from a state s to a state s' by executing the joint action $\overrightarrow{a}$.

A sailing match race is modeled as a stochastic game. The set of players is limited to two players, the actions available to a yacht, that is a player, are limited to tack decisions taken according to wind and the opponent's behavior. The reward functions are given by negative, null or positive values (i.e., $-1, 0, 1$). Actors in a game are the players whose intents are to either maximize gains or minimize losses. In order to model the match racing as a game, a player represents the skipper and his yacht. Therefore, a sailing match racing is a game with two players ($n = 2$). b_i denotes a player where $i = \{1, 2\}$. During the race, the actions available to a player are changing the yacht's tack from port to starboard and vice versa. At a given speed, each player b_i has only two possible actions: **to tack** or **not to tack**. Let us consider for example yachts A and B where A is leading and B is trailing (Fig. 5). In an upwind leg, to achieve its goal, yacht A reacts to yacht B behavior:

- B on the same tack; A continues until the layline and tacks once to limit the tacking cost (Fig. 5(a)).
- B on the opposite tack; A tacks to minimize risk. In this case, A achieves one more tack but minimizes the risk by a loose covering (Fig. 5(b)).

A tack action forces the yacht to turn through the wind and slow down. We assume a known speed loss for tacking (v_{totack}). A_i denotes the set of actions for each yacht b_i, $(A_i)_{i\in\{1,2\}} = \{$to tack, not to tack$\}$. $\overrightarrow{a} = (a_1, a_2)$ denotes the joint action of players b_1 and b_2 ($\overrightarrow{a} \in \{\overrightarrow{a}_{(t,t)}$=(tack, tack), $\overrightarrow{a}_{(t,n)}$=(tack, not to tack), $\overrightarrow{a}_{(n,t)}$=(not to tack, tack), $\overrightarrow{a}_{(n,n)}$=(not to tack, not to tack)$\}$). Each game state is given by the two yacht spatial states and the wind direction (wind speed is considered as homogeneous). Formally, a state s is a tuple $s = \langle Y_1, Y_2, W\rangle$:

- A yacht spatial state (Y_i) denotes the yacht location, direction and speed. Formally $Y_i = \langle pos, \overrightarrow{v}\rangle$ where (1) $pos = \langle x, y\rangle$ denotes the current yacht location and (2) $\overrightarrow{v} = \langle \theta, v\rangle$ denotes the current yacht direction and speed;

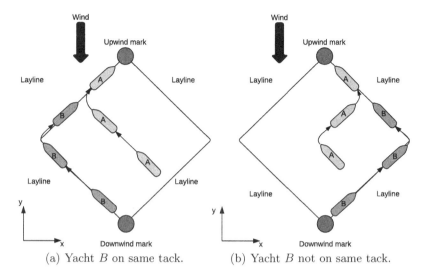

(a) Yacht B on same tack.   (b) Yacht B not on same tack.

Fig. 5. Yacht A behavior according to the opponent's spatial state

- Wind direction $W \in \{-45°, -40°, \ldots, 0°, \ldots, 45°\} \times \mathbb{R}^+$, where $0°$ represents the direction to the upwind mark [6]. The other states represent shifts of $\pm5°$ from that direction.

A finite state space represents the environment upon which the game is played while transiting from one state to another is a stochastic process. The match racing game consists in a finite set of states $S = \{s^0, s^1, \ldots, s^h\}$. We assume complete observability, that is, at current time, each yacht b_i has a complete knowledge about the opponent spatial state and the wind direction. The final state $s^h \in S$ is given when the first yacht reaches the upwind mark. During the race the challenge for yacht b_1 is to stay ahead the opponent b_2. We define this advantage as the estimated time to reach the upwind mark from the current location, and is evaluated according to a straight weather routing derivation. During the race, when a yacht b_1 makes a decision to tack or not to tack, the remaining time to reach the upwind mark is evaluated. More formally, let us define for each yacht a function denoted $\Gamma_i(s)$ that estimates the time to reach the upwind mark (UM). This function depends on the current yacht location relative to the upwind mark $(dist(pos^{UM}, Y_i.pos))$, where; (1) Y_i denotes the current state (s) of yacht b_i and (2) $pos^{UM} = (x^{UM}, y^{UM})$ denotes the location of the upwind mark. At each state s, according to $\Gamma_i(s)$, three cases can be distinguished in order to derive which of the two yachts is the leader:

- $\Gamma_1(s) < \Gamma_2(s)$: yacht b_1 takes less time than the opponent b_2 to reach the upwind mark, then yacht b_1 is the leader.
- $\Gamma_1(s) > \Gamma_2(s)$ yacht b_1 takes more time than the opponent b_2 to reach the upwind mark, then yacht b_1 is the follower.
- $\Gamma_1(s) = \Gamma_2(s)$: both yachts are to a tie.

Accordingly the payoff function $p_i(s)$ is defined for each state as follows:

$$p_1(s) = \begin{cases} 1 \Leftrightarrow \Gamma_1(s) < \Gamma_2(s) & \text{if } b_1 \text{ is the leader,} \\ -1 \Leftrightarrow \Gamma_1(s) > \Gamma_2(s) & \text{if } b_1 \text{ is the follower,} \\ 0 \Leftrightarrow \Gamma_1(s) = \Gamma_2(s) & \text{otherwise.} \end{cases}$$

In order to win the race, a yacht b_i must reach the upwind mark first. We set the reward for the goal state to either positive or negative values. The reward function R_i is defined for each yacht as follows:

$$R_i = \begin{cases} 1 & \text{if } b_i \text{ wins} \\ -1 & \text{if } b_i \text{ loses} \\ 0 & \text{otherwise.} \end{cases}$$

At each step k, a two-dimensional matrix game G^k represents the state s^k (i.e., at each state the player is faced to a game stage). The matrix game G^k gives the payoffs for b_1 and b_2. The matrix game is defined as follows: one player b_1 denotes the rows and the other b_2 the columns. The entries in the matrix are the payoffs to the row players. An important property of this game is $p_1(s) = -p_2(s)$ which means that a payoff for one player is a cost for the other.

$$G^k = \begin{array}{cc} & \begin{array}{cc} \text{to tack} & \text{not to tack} \end{array} \\ \begin{array}{c} \text{to tack} \\ \text{not to tack} \end{array} & \left(\begin{array}{cc} p_1(s^k), & p_1(s^k) \\ p_1(s^k), & p_1(s^k) \end{array} \right) \end{array}$$

At each state s^k, each player b_i has a finite set of actions at his disposal to choose from $(A_i)_{i \in \{1,2\}} = \{\text{to tack, not to tack}\}$. Therefore, the resulting bi-matrix G^k therefore is the matrix game for both players. When acting, two players transit from one state s^k to another state s^{k+1} leading to a transition from one game stage to another one. During a race, each yacht makes progress towards the upwind mark with stochastically varying wind and according to the sailing actions taken. The wind evolution is subject to some uncertainties. The transition function captures the stochastic nature of the yacht's motion. The game transits from one state to another according to a probability distribution. The state transition probability is a function of both the players' actions, and the current state. Formally, the transition function $T(s^k, \vec{a}, s^{k+1}) = Pr(s^{k+1}|s^k, \vec{a})$ is the probability that the system transits to state s^{k+1} when the joint action $\vec{a}$ is taken at state s^k. The transition function is based on the predicted yacht locations. The transition function $T(s^k, \vec{a}, s^{k+1})$ corresponds to the probability that the wind shifts $(Pr(W^{k+1}|W^k)$. Let $s^k = \langle Y_1^k, Y_2^k, W^k \rangle$ and $s^{k+1} = \langle Y_1^{k+1}, Y_2^{k+1}, W^{k+1} \rangle$ two distinct states. Moreover, we assume that the wind and the yacht locations are independent so the wind is not affected by any joint action taken $\vec{a}$. Moreover, we assume that the joint action $\vec{a}$ is deterministic. Therefore:

$$
\begin{aligned}
T(s^k, \overrightarrow{a}, s^{k+1}) &= Pr(s^{k+1}|s^k, \overrightarrow{a}) \\
&= Pr(Y_1^{k+1}, Y_2^{k+1}, W^{k+1}|Y_1^k, Y_2^k, W^k, \overrightarrow{a}) \\
&= Pr((Y_1^{k+1}, Y_2^{k+1})|(Y_1^k, Y_2^k), \overrightarrow{a}) * Pr(W^{k+1}|W^k, \overrightarrow{a}) \\
&= Pr(W^{k+1}|W^k)
\end{aligned}
\tag{1}
$$

Let us define how a state s^{k+1} is computed from a state s^k and a joint action $\overrightarrow{a}$. Let us consider two states $s^{k+1} = \langle Y_1^{k+1}, Y_2^{k+1}, W^{k+1} \rangle$ and $s^k = \langle Y_1^k, Y_2^k, W^k \rangle$ and the joint action $\overrightarrow{a}$. For each yacht b_i and action a_i, Y_i^{k+1} is computed at each time step as follows:

$$
Y_i^{k+1} =
\begin{cases}
x^{k+1} = v^{k+1} * \delta t * sin(\theta^{k+1}) + x^k \\
y^{k+1} = v^{k+1} * \delta t * cos(\theta^{k+1}) + y^k \\
\theta^{k+1} = \begin{cases} -\theta^k + shift & \text{if } a_i = \text{tack} \\ \theta^k + shift & \text{if } a_i = \text{not to tack.} \end{cases} \\
v_i^{k+1} = \begin{cases} v_i^k - v_{totack} & \text{if } a_i = \text{to tack} \\ v & \text{if } a_i = \text{not to tack.} \end{cases}
\end{cases}
$$

During the race and at each time step, each yacht should decide to tack or to not tack. v_{totack} denotes a known value corresponding to the speed loss associated to the tacking action. A possible transition from the game stage G^k to another game stage G^{k+1} depends on the outcome of G^k and the joint action given by the matrix game G^k. The probability of transiting from a given game stage G^k when acting with a joint action $\overrightarrow{a}$ is the probability to transit from one state S_{G^k} to a state $S_{G^{k+1}}$. Players simultaneously choose a row and a column of the matrix game causing player b_1 to for instance win the payoff $p_1(s^k)$ from player b_2 who loses the same amount. Therefore, the game moves to another stage, with a probability that depends on the selected joint action and the current stage.

Formally, for each state s^k; $Pr(G^k, \overrightarrow{a}, G^{k+1}) = \sum_{s^{k+1} \in S_{G^{k+1}}} T(s^k, \overrightarrow{a}, s^{k+1})$. To compute $Pr(G^k, \overrightarrow{a}, G^{k+1})$, let us consider the state s^k and two possible joint actions $\overrightarrow{a}_0$ and $\overrightarrow{a}_1$ (Fig. 6). By applying the joint action $\overrightarrow{a}_0$ to the state s^k, there are two possible states s_1^{k+1} and s_2^{k+1} $(T(s^k, \overrightarrow{a}_0, s_1^{k+1}), T(s^k, \overrightarrow{a}_0, s_2^{k+1}))$. By applying the joint action $\overrightarrow{a}_1$ to the state s^k, there are two possible states s_3^{k+1} and s_4^{k+1} $(T(s^k, \overrightarrow{a}_1, s_3^{k+1}), T(s^k, \overrightarrow{a}_1, s_4^{k+1}))$. These states $s_1^{k+1}, s_2^{k+1}, s_3^{k+1}$ and s_4^{k+1} define the same matrix game G_1^{k+1}. Therefore:

$$
\begin{aligned}
Pr(G^k, \overrightarrow{a}, G_1^{k+1}) = T(s^k, \overrightarrow{a}_0, s_1^{k+1}) + T(s^k, \overrightarrow{a}_0, s_2^{k+1}) \\
+ T(s^k, \overrightarrow{a}_1, s_3^{k+1}) + T(s^k, \overrightarrow{a}_1, s_4^{k+1})
\end{aligned}
$$

At a given state s^k the matrix game G^k is assigned to a value $V(k)$. For each matrix game, the game value $V(k)$ is the unique solution of G^k. A long-term strategy should maximize the gain expectation. Accordingly, we introduce a value function for each game G^k at each state s^k, as follows:

$$
V(k) = \max_{\overrightarrow{a}}
\begin{cases}
\sum_{s^k, G^{k+1}} Pr(G^k, \overrightarrow{a}, G^{k+1})V(k+1) & k < h \\
R_i & k = h.
\end{cases}
$$

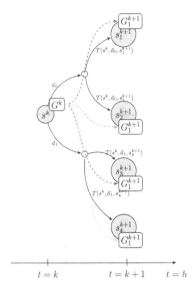

Fig. 6. Transitions between states and games

$Pr(G^k, \overrightarrow{a}, G^{k+1})$ denotes the probability of transiting from game G^k when acting with a joint action $\overrightarrow{a}$. A value iteration algorithm is applied to maximize the expected rewards as suggested in [10]. There is a mixed strategy for player b_1 such that b_1's average gain is at least V no matter what b_2 does and there is a mixed strategy for player b_2 such that b_2's average loss is at most V no matter what b_1 does. Also, if $V = 0$, the game is neutral. If $V > 0$ the game is said to be favorable to player b_1, otherwise if $V < 0$ the game favors player b_2. The row player tries to match the column player while the column player tries to guess the opposite of the row player. The value of the game can be derived as either the minimum of what the row player can achieve knowing the strategy of the column player or the maximum of what the column player can hold the row player to, knowing the strategy of the row player. In order to derive the solution of the game, the algorithms have to recursively iterate over each game stage.

4   Sailing Match Race: Case Study

We introduce an example of a sailing yacht match race with two yachts b_1 and b_2. Let us initialize the set of actions available to each player as follows: $(A_i)_{i \in \{1,2\}} = \{\text{to tack, not to tack}\}$. We assume the following initial state:

$$s^0 = \begin{cases} Y_1 = \langle 15, 130, -45°, 13 \rangle \\ Y_2 = \langle 25, 130, +45°, 13 \rangle \\ W = 0° \end{cases}$$

b_1 is sailing on starboard tack and is located to the left of b_2 sailing on port tack. Both yachts have to travel exactly the same distance in the wind direction to reach the upwind mark. The yachts speed is $v = 13kt$ and the speed loss when tacking is $3kt$. The wind speed is considered as constant. We consider a finite horizon h ($h = 10$, $\delta t = 30s$)), where the wind shifts to the left ($shift = -5°$) twice, at times $k = 4$ and $k = 8$. The wind direction signal is sampled every three seconds, and the corresponding wind directions are placed in bins of amplitude $5°$ ($shift = \{-5, 0, +5\}$). Let us focus on the b_1 behavior. b_1 is the row player while b_2 is the column player. $p_1(s^k)$ denotes the payoff for player b_1 in the state s^k. s^h denotes the final state in which yacht b_1 reaches first the upwind mark.

$$
G_1^1 = \begin{array}{cc} & \begin{array}{cc} \text{to tack} & \text{not to tack} \end{array} \\ \begin{array}{c} \text{to tack} \\ \text{not to tack} \end{array} & \left(\begin{array}{cc} 0, & -1 \\ 1, & 0 \end{array} \right) \end{array}
$$

At step $k = 1$, the wind direction remains the same ($W^1 = 0°$). There are four possible states ($s_1^1, s_2^1, s_3^1, s_4^1$) and one possible matrix game (G_1^1) (Fig. 7).

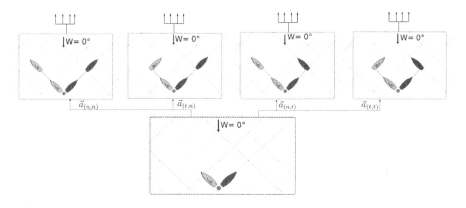

Fig. 7. Decision tree: spatial representation

According to Fig. 7, there are four possible states ($s_1^1, s_2^1, s_3^1, s_4^1$) denoted respectively as $\overrightarrow{a}_{(n,n)}$, $\overrightarrow{a}_{(n,t)}$, $\overrightarrow{a}_{(t,n)}$ and $\overrightarrow{a}_{(t,t)}$:

- if yachts b_1 and b_2 do not tack ($\overrightarrow{a}_{(n,n)}$), there is no leader ($s_4^1$) at step $k = 1$;

$$
s_1^1 = \begin{cases} Y_{b_1} = \langle -126.87, 271.87, -45°, 13 \rangle \\ Y_{b_2} = \langle 166.87, 271.87, +45°, 13 \rangle \\ W = 0°. \end{cases}
$$

- if yacht b_1 doesn't tack and yacht b_2 tacks ($\overrightarrow{a}_{(n,t)}$), b_1 is the leader at step $k = 1$;

$$
s_2^1 = \begin{cases} Y_{b_1} = \langle -126.87, 271.87, -45°, 13 \rangle \\ Y_{b_2} = \langle -84.20, 239.20, -45°, 10 \rangle \\ W = 0°. \end{cases}
$$

– if yacht b_1 tacks and yacht b_2 doesn't tack $(\overrightarrow{a}_{(t,n)})$, b_2 is the leader at step $k = 1$;

$$s_3^1 = \begin{cases} Y_{b_1} = \langle 124.20, 239.20, +45°, 10 \rangle \\ Y_{b_2} = \langle 166.87, 271.87, +45°, 13 \rangle \\ W = 0°. \end{cases}$$

– if both yachts b_1 and b_2 tack $(\overrightarrow{a}_{(t,t)})$, there is no leader at step $k = 1$;

$$s_4^1 = \begin{cases} Y_{b_1} = \langle 124.20, 239.20, +45°, 10 \rangle \\ Y_{b_2} = \langle -84.20, 239.20, -45°, 10 \rangle \\ W = 0°. \end{cases}$$

According to G_1^1, there is only one situation which is favorable to yacht b_1. At step $k = 2$ the wind direction remains the same. Considering the 4 situations at step $k = 1$ $(s_1^1, s_2^1, s_3^1, s_4^1)$ and the 4 joint actions $(\overrightarrow{a}_{(n,n)}, \overrightarrow{a}_{(n,t)}, \overrightarrow{a}_{(t,n)}$ and $\overrightarrow{a}_{(t,t)})$, there are 16 possible states $(s_1^2, \ldots, s_{16}^2)$ and 3 possible matrices games (G_1^2, G_2^2, G_3^2) at step $k = 2$ (Table 1):

– G_1^2 means:
  • if yacht b_1 tacks and yacht b_2 tacks, there is no leader at the next step $k = 2$ (the payoff of yacht b_1 is 0);
  • if yacht b_1 tacks and yacht b_2 does not tack, yacht b_2 is the leader at the next step $k = 2$ (the payoff of yacht b_1 is -1);
  • if yacht b_1 does not tack and yacht b_2 tacks , yacht b_1 is the leader at the next step $k = 2$ (the payoff of yacht b_1 is 1);
  • if yacht b_1 does not tack and yacht b_2 does not tack, there is no leader at the next step $k = 2$ (the payoff of yacht b_1 is 0);
– G_2^2 means:
  • yacht b_2 is the leader at the next step $k = 2$ (the payoff of yacht b_1 is -1, Table 1), if both yachts b_1 and b_2 tack or not, or if yacht b_1 tacks and yacht b_2 does not tack;
  • if yacht b_1 does not tack and yacht b_2 tacks, there is no leader at the next step $k = 2$ (the payoff of yacht b_1 is 0);
– G_3^2 means:
  • yacht b_1 is the leader at the next step $k = 2$ (the payoff of yacht b_1 is 1), if both yachts b_1 and b_2 tack or not, or if yacht b_1 does not tack and yacht b_2 tacks;
  • if yacht b_1 tacks and yacht b_2 does not tack, there is no leader at the next step $k = 2$ (the payoff of yacht b_1 is 0).

At each time step k, similar operations are derived in order to compute the states of the next step $k + 1$ according to the current state, joint action and possible wind shifts. This supports a derivation of the matrix games at each step and can be represented as a decision tree as illustrated by Fig. 8 that depicts a part of the global decision tree as applied to our example. A winning strategy for yacht b_1 is to find a path in the global decision tree from s^0 to s^h. There is one final state s^h in which yacht b_1 is the winner. The decision tree corresponding to

Table 1. Matrices games at step $k = 2$ according to joint actions.

$k = 1$	Actions and States ($k = 2$)		Matrix Game ($k = 2$)
s_1^1, G_1^1	$\vec{a}_{(n,n)}$	s_1^2	$G_1^2 = \begin{matrix} \text{to tack} \\ \text{not to tack} \end{matrix} \begin{pmatrix} \text{to tack} & \text{not to tack} \\ 0 & -1 \\ 1 & 0 \end{pmatrix}$
	$\vec{a}_{(n,t)}$	s_2^2	
	$\vec{a}_{(t,n)}$	s_3^2	
	$\vec{a}_{(t,t)}$	s_4^2	
s_2^1, G_1^1	$a_{(n,n)}$	s_5^2	$G_2^2 = \begin{matrix} \text{to tack} \\ \text{not to tack} \end{matrix} \begin{pmatrix} \text{to tack} & \text{not to tack} \\ -1 & -1 \\ 0 & -1 \end{pmatrix}$
	$\vec{a}_{(n,t)}$	s_6^2	
	$\vec{a}_{(t,n)}$	s_7^2	
	$\vec{a}_{(t,t)}$	s_8^2	
s_3^1, G_1^1	$\vec{a}_{(n,n)}$	s_9^2	$G_3^2 = \begin{matrix} \text{to tack} \\ \text{not to tack} \end{matrix} \begin{pmatrix} \text{to tack} & \text{not to tack} \\ 1 & 0 \\ 1 & 1 \end{pmatrix}$
	$\vec{a}_{(n,t)}$	s_{10}^2	
	$\vec{a}_{(t,n)}$	s_{11}^2	
	$\vec{a}_{(t,t)}$	s_{12}^2	
s_4^1, G_1^1	$\vec{a}_{(n,n)}$	s_{13}^2	$G_1^2 = \begin{matrix} \text{to tack} \\ \text{not to tack} \end{matrix} \begin{pmatrix} \text{to tack} & \text{not to tack} \\ 0 & -1 \\ 1 & 0 \end{pmatrix}$
	$\vec{a}_{(n,t)}$	s_{14}^2	
	$\vec{a}_{(t,n)}$	s_{15}^2	
	$\vec{a}_{(t,t)}$	s_{16}^2	

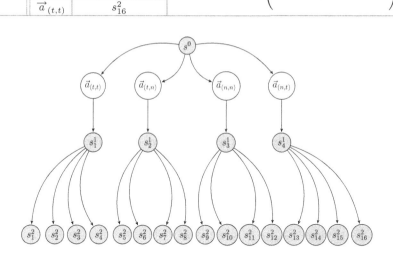

Fig. 8. Decision tree

this scenario includes a series of possible paths between s^0 and s^h. These paths define the winning strategies available for yacht b_1. More formally, computing a winning strategy implies to solve the value function and where at each step k, $V(k)$ is recursively computed up to the final state h. The computed value assigns to each state the maximum expected value expressing the probability to win, and considers for this state the action maximizing this probability. Regarding the considered scenario, there are 4^{10} possible states, while among the $4^3 = 64$

possible matrix games, only 46 distinct matrices games arise. Let us study the yacht b_1 behavior, where his aim is to win the race regarding the initial state s^0, yacht b_2 behavior and wind fluctuations during the race. A possible winning strategy is hereafter illustrated (Fig. 9).

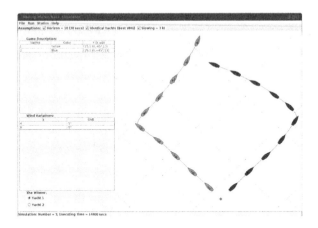

Fig. 9. Winning strategy for yacht b_1

Figure 9 illustrates the principles of the interface developed so far for the implementation of the stochastic game. During the first part of the race, at steps $k = 0$ up to $k = 3$, with an initial wind ($W = 0°$), both yachts b_1 and b_2 do not tack. Therefore, there is no leader at time step $k = 4$. At step $k = 4$ the wind shifts to the left ($shift = -5°$). Again, both yachts b_1 and b_2 do not tack. Yacht b_1 becomes the leader as his course is closer to the direction of the upwind mark. At time step $k = 5$, the wind is to the left ($W = -5°$), both yachts b_1 and b_2 tack. Yacht b_2 remains the leader at step $k = 6$ but the gain towards the mark is higher for b_1. From this step, yacht b_1 takes the lead. Indeed, the wind direction is more favorable to b_1 than to b_2. This scenario is reflected by the following matrices games. During the first part of the race, at steps from $k = 0$ up to $k = 4$, the matrix game is as follows:

$$
G_1^1 = G_1^2 = G_1^3 = G_1^4 = \begin{array}{c} \\ \text{to tack} \\ \text{not to tack} \end{array} \overset{\begin{array}{cc} \text{to tack} & \text{not to tack} \end{array}}{\left(\begin{array}{ccc} 0 & , & -1 \\ 1 & , & 0 \end{array} \right)}
$$

According to the matrices games $G_1^1 = G_1^2 = G_1^3 = G_1^4$ and Fig. 10, both yachts b_1 and b_2 have decided to not tack. At time steps $k = 5$ and $k = 6$, the wind is to the left ($W = -5°$). Whatever joint action, yacht b_1 will be the leader. This is reflected by the following matrices games such as:

$$
G_1^5 = G_1^6 = \begin{array}{c} \\ \text{to tack} \\ \text{not to tack} \end{array} \overset{\begin{array}{cc} \text{to tack} & \text{not to tack} \end{array}}{\left(\begin{array}{ccc} -1 & , & -1 \\ -1 & , & -1 \end{array} \right)}
$$

At step $k = 7$, as long as yacht b_1 does not tack, she is the leader:

$$G_1^7 = \begin{array}{c} \\ \text{to tack} \\ \text{not to tack} \end{array} \begin{array}{cc} \text{to tack} & \text{not to tack} \\ \left(\begin{array}{ccc} 0 & , & -1 \\ 1 & , & 1 \end{array} \right) \end{array}$$

At steps from $k = 8$ up to $k = 10$, for any considered joint action, yacht b_1 remains the leader:

$$G_1^8 = G_1^9 = G_1^{10} = \begin{array}{c} \\ \text{to tack} \\ \text{not to tack} \end{array} \begin{array}{cc} \text{to tack} & \text{not to tack} \\ \left(\begin{array}{ccc} 1 & , & 1 \\ 1 & , & 1 \end{array} \right) \end{array}$$

This scenario outcome is consistent with usual sailing strategies. For instance, if the wind shifts to one side during an upwind leg, the yacht positioned to that side takes an advantage. This shows how the wind and the opponent's actions are taken into account. According to Fig. 10, despite different behaviors of yacht b_2, yacht b_1 remains the winner. This is consistent with real races. Indeed, when the wind is continuously shifting to the one side during the upwind leg, the optimal route includes one tack and starts going to this side. The experiments show that the model and strategic game developed so far comply relatively with common sailing crew behaviors. The scenarios developed cannot reflect the whole complexity of yacht race configurations and crew decisions, but they provide a useful framework for debriefing and training purpose. An important challenge sailors face in yach racing strategy is to foresee the wind evolution to take advantage of subsequent wind fluctuations. However, as the wind is often uncertain, the leader should cover the oponent by minimizing the lateral distance between both yachts, in order to prevent the trailing yacht to obtain an advantage from an unforseen wind shift (Sect. 2). These two distinct objectives -weather routing and covering the opponent- may lead to opposite decisions. It is one of the most difficult tasks to choose from. Overall, the modeling approach can simulate a series of configurations and then help to identify some key behaviors for a winning strategy. One of the advantages of the modeling approach is that it can give useful insights to match racing strategy for configurations where both yachts are not too close to each other.

5   Related Work

Most of related works on yacht racing have been oriented to the modeling of yacht performance, mostly by developing Velocity Prediction Programs [8]. Weather routing models the wind as a stochastic process represented as a Markov chain (MDP) [2,3,5]. In [13] a method for short term weather routing for a small robotic boat is introduced. It assumes fixed wind conditions, the objective being to reach a given target position, but this does not guarantee an optimal trajectory regarding the minimal time to reach the goal as expected in competitive sailing. In [5], a *Velocity Prediction Program* (VPP) predicts a yacht speed and direction

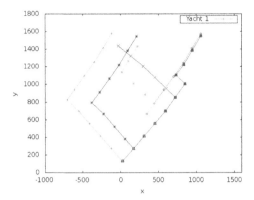

Fig. 10. Winning strategy for yacht b_1 (1) despite different behaviors of yacht b_2

of sail given a wind speed and direction as well as the yacht's wind angle. This VPP is associated to a MDP wind variation to compute a race. A similar work is presented in [12] and that models the decision-making problem as a MDP but assumes yachts dynamics as deterministic and only considers wind variations. In [1], the authors introduced a yacht race stochastic model that integrates wind variations, but without taking into account different navigation strategies. Overall, most of these studies are oriented to weather routing under weather variations but fixed strategy, while any opponent behavior is not completely taken into account. In [1], minimizing the distance to the opponent is also taken into account in the leading yacht strategy. To the best of our knowledge, a few studies addressed yachts interactions during a race. Philpott et al. [1] studied the effect on the yacht performance of the proximity of another yacht and the impact to the wind field. Recently, Tagliaferri et al. [6] introduced a risk attitude model and showed that a risk management policy that depends on the yachts relative locations is better than a straight strategy seeking to minimize the time to reach the finish line: the leading yacht should be risk averse while the trailing yacht should seek risk. However, the opponent actions and lateral locations on the race course are not considered and the risk attitude is based on wind shifts only. Nevertheless, one of the strategic basics in match racing is that the leading yacht should most often cover his opponent or at least minimize the lateral distance between both yachts, so should react to the opponent actions.

6   Conclusion

This paper introduces an action-based approach for modeling a yacht match race under uncertain wind. A stochastic game between two non-cooperative players is associated to a stochastic wind MDP model that takes into account wind fluctuations and yachts' reactions to their opponent's location and actions. A sailing match race consists of two similar yachts racing against each other. A winning strategy based on a dynamic and uncertain wind and opponent's behavior is developed. The stochastic game considers discrete states and actions of two non-cooperative players. The objective is to allow a skipper during the race to make decisions that maximizes his likelihood to win. Preliminary experiments not only can either reflect some observed behaviors but also favor discussions amongst the skippers to analyze some of the decisions taken and resulting outcomes. Winning strategies can be simulated as well as key behaviors during a race. The applicability and realism of our model is still to be extended by additional constraints such as local wind fluctuations and wind interactions with the other competitor in the vicinity as well as navigation rules. Those are left to further work. We also plan to apply the modeling approach to real world races as well as different leg configurations.

References

1. Philpott, A., Henderson, S.G., Teirney, D.: A simulation model for predicting yacht match race outcomes. Oper. Res. **52**(1), 1–16 (2004). INFORMS
2. Philpott, A.: Stochastic optimization and yacht racing. Appl. Stochast. Program. **5**, 315–336 (2005)
3. Dalang, C.R., Dumas, F., Sardy, S., Morgenthaler, S., Vila, J.: Stochastic optimization of sailing trajectories in an upwind regatta. J. Oper. Res. Soc. (2014). Palgrave Macmillan
4. Martin, D., Beck, R.: PCSAIL, a velocity prediction program for a home computer. In: The 15th Cheasapeake Sailing Yacht Symposium, vol. 100 (2001)
5. Ferguson, D.S., Elinas, P.: A Markov decision process model for strategic decision making in sailboat racing. In: Butz, C., Lingras, P. (eds.) AI 2011. LNCS (LNAI), vol. 6657, pp. 110–121. Springer, Heidelberg (2011). https://doi.org/10.1007/978-3-642-21043-3_14
6. Tagliaferri, F., Philpott, A.B., Viola, I.M., Flay, R.G.J.: On risk attitude and optimal yacht racing tactics. J. Ocean Eng. (2014)
7. White, J.D.: A survey of applications of Markov decision processes. J. Oper. Res. Soc. JSTOR, 1073–1096 (1993)
8. Kerwin, J.E.: A Velocity Prediction Program for Ocean Racing Yachts. Massachusetts Institute of Technology, Department of Ocean Engineering (1978)
9. Shapley, L.S.: Stochastic games. Proc. Natl. Acad. Sci. USA **93**(10), 1095 (1953)
10. Littman, M.L.: Markov games as a framework for multi-agent reinforcement learning. In: ICML 1994, pp. 157–163 (1994)
11. Matos, M., Cristina, P., Ferreira, M.A.M.: Game theory, the science of strategy. Int. J. Latest Trends Financ. Econ. Sci. **4**(2), 738 (2014)
12. Roncin, K., Kobus, J.M.: Dynamic simulation of two sailing boats in match racing. Sports Eng. **7**(3), 139–152 (2004)
13. Stelzer, R., Proll, T.: Autonomous sailboat navigation for short course racing. Robot. Auton. Syst. **56**(7), 604–614 (2008)
14. Richards, P.J., Le Pelley, D.J., Jowett, D., Little, J., Detlefsen, O.: A wind tunnel study of the interaction between two sailing yachts. In: 21st Chesapeake Sailing Yacht Symposium, Annapolis, Maryland, pp. 162–168 (2013)
15. Richards, P.J., Aubin, N., Le Pelley, D.J.: The interaction between sailing yachts in fleet and match racing situations. In: 23rd International HISWA Symposium, Amsterdam (2014)

Towards a Modelling and Optimisation of the Recovery of Marine Floating Plastic

Nicolas Maslov[1]([✉]), Loïc Salmon[2], and Christophe Claramunt[1]

[1] Naval Academy Research Institute, 29160 Lanvéoc, France
nico.maslov@gmail.com
[2] Université de Bretagne Occidentale, 29200 Brest, France

Abstract. The recovery of marine plastic litter has been recently recognised as one of the world greatest environmental challenge. While many recent initiatives are oriented towards the recovery of micro plastics, there is still a need for the development of observation and in situ capabilities that will provide appropriate positional data to combat such critical pollution phenomenon. The experimental research developed in this paper introduces a methodology for geo-locating and collecting plastic wastes in a maritime environment. The approach is spatially and temporally-based in the sense plastic wastes are tracked over time according to some meteorological conditions (i.e., winds, currents, tides). Overall, predicting plastic waste displacements constitutes a first step to an optimization process whose objective is to collect plastic debris in littoral zones using a dedicated ship.

Keywords: Marine Litter · Macro-plastic · Trajectory prediction · Route optimisation

1 Introduction

Plastic pollution is considered as a major environmental problem [1], with three trillion of plastic wastes populating the oceans, weighting about 322 million tons[2]. This growing accumulation in seas and oceans constitutes a major risk for the marine ecosystems [3], particularly because of its introduction into the food chain. Furthermore, it has been observed that anthropogenic marine debris may cause physical harms to humans when debris is ingested via seafood [4]. Last, marine plastic pollution also has a significant impact on tourism in coastal environments. Plastics are mainly produced from substances extracted from oil and natural gas. First highlighted for their remarkable mechanical properties, these materials are now suffering from the problematic of their end of life due to their slow degradation. During the wearing process, plastics fragment, spread in nature and take 500 to 1000 years to decompose [5].

Plastic waste can be categorised according to their dimensions, and into three types: macro-plastic, meso-plastic and micro-plastic [6], EU [7], NOAA [8]. A distinction can be made between micro-plastics for plastic debris sizing less

© Springer Nature Switzerland AG 2020
S. Di Martino et al. (Eds.): W2GIS 2020, LNCS 12473, pp. 214–229, 2020.
https://doi.org/10.1007/978-3-030-60952-8_21

than 5 mm, meso-plastics for dimensions between 5 mm and 2.5 cm and macro-plastics for plastic debris larger than 2.5 mm. However, it has been observed that researches dealing with marine plastic pollution mainly concerns micro-wastes. [9,10].

The objective of our preliminary research is to introduce a methodology to first observe, model and map macro-plastic phenomena at sea, and that will facilitate the recovery of large floating macro-plastic areas in coastal marine environments. Macro-plastics are generally large plastic elements such as bags, water bottles or fishing equipment. The main idea behind our study is to combine different observation resources (i.e., high resolution satellite data) with in situ observations in order to provide a decision-aid mechanism for a recovery ship whose objective will be to collect and recover large plastic areas in coastal environments. Such large plastic patches are generated by convergence of oceanographic currents and meteorological conditions in the ocean and which have been already observed and studied [11–13]. Indeed, coastal areas are appropriate targets, especially close to river mouths, as it has been observed that each year rivers discharge about 2 million tons of plastic waste [5].

Before being collected, macro-plastic should be identified and monitored. So far two main means of observation can be operated such as remote sensing techniques using conventional or radar images or in situ observations. However, and to the best of our knowledge, still not completely convincing approaches have been developed to successfully observe and track macro-plastic areas over time in marine areas. This motivates our search for an integrated approach, based on a combination of high-resolution satellite data with in situ observations. A data modelling and optimisation framework is developed, and whose objective is to track macro-plastic areas overs time and to visualise them in a dynamic cartographical environment. The final objective is to optimise a ship trajectory to collect those plastic macro-waste taking into account their drift.

The remain of this article is organized as follows. Section 2 introduces the data acquisition sources and provides details of a marine observation protocol. Section 3 presents extracting and monitoring approach and the underlying database model. Section 4 develops two drift models depending on marine currents applied to floating waste, the first one proposed is deterministic while the second one is stochastic to handle conditions uncertainty and variability at sea, the objective being to optimise ship trajectories when recovering marine floating plastic. Section 5 presents a case study that illustrates the collection model while Sect. 6 concludes the paper and outlines future developments.

2   Marine Waste Patches Observation and Recovery

The main principles of the derivation of an optimal route to efficiently collect plastic waste at sea are described in the following Fig. 1. This requires first of all (1) a data model for the recovery and management of waste patch geo-located data. Secondly (2) the data model must take into account fusion and update aspects while different observation sources are available and waste patch are constantly moving due to currents and winds. (3) An extraction of a plastic-waste density map moving according with a drift model of the waste patch. This process should also be involved in the regular update an underlying database. Finally (4) a decision support algorithm to derive the most favourable route to collect as much as possible plastic wastes while taking into account the limited range of a vessel. A series of contextual parameters will be taken into account such as the amount of collected plastic, the time of sailing or the energy consumed by the boat (i.e., an optimised route would avoid currents and bad weather conditions).

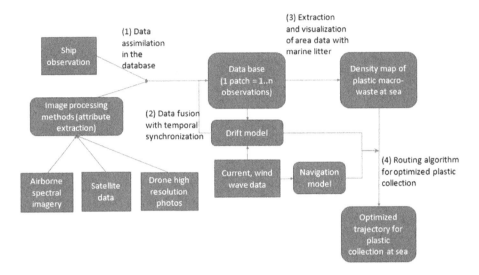

Fig. 1. Observing and collecting macro-plastic wastes at sea: main principles

It clearly appears that a combination of complementary observation methods is necessary in order to model and track waste patches at sea [14]. Indeed, each observation technology provides a specific information mean to be recovered at different resolutions (as well as at variable costs). In the case of plastic waste at sea, aerial or satellite images allow, for example, a relative evaluation of the location and estimation of the surface area of a floating waste patch. While the temporal resolution is limited by the orbit characteristics for satellite images and the availability of satellite images, for airborne systems a data acquisition implies the planning of an observation campaign this being not always easy and probably a relatively costly operation. Moreover, satellite orbits considerably vary considerably (i.e., from a few days to 1 month); aircraft are generally limited to about ten flight hours, and about thirty hours for drones. Weather constraints are also a constraint for all optical observation instruments but less so when it comes to microwaves [14]. Due to many different types and size of plastics, solutions based on satellite and airborne remote sensing tools for assessing ocean plastic pollution are not completely efficient when applied alone. However, remote sensing technologies with moderate to high temporal, spectral and spatial resolutions have the potential to be a reliable source of quantitative and qualitative information at wide geographical scale [15]. In [16], a first step towards floating marine macro plastics using hyperspectral remote sensing algorithm is developed, due to the unique spectral signature of polymers in near-infrared part of electromagnetic spectrum. High resolution photo have been used in coastal detection in [17,18] for instance. Overall, and as a complete temporal availability of high resolution images is not a guarantee, complementary in situ additional observation data is required. Last, and as tracking plastic waste areas is a key issue for recovery missions, such integrated observations should be therefore and ideally coupled with oceanographic drift models.

An efficient way to collect detailed information on the characteristics of floating waste is by in situ observation surveys. Recently, a data acquisition protocol for marine macro-plastic collection and waste-to-energy transformation has been suggested[1]. Voluntary observations come from crews on ships. The surveys are in paper format but a willing to develop a numerical format is under consideration, with the objective to provide an open database. Three observation situations of floating waste patches are defined according to the possibilities of the crew and the vessel. Considering the speed of the boat and the willing of people on board to report information, an observation can be a rapid visual estimation of the waste, by completing a general observation sheet and taking photos or videos in the basic case. this gives a succession of GPS-based location in a more refined case with information about distance between the boat and the waste. Finally, some other information can be provided concerning the waste composition.

PLASTIC WASTE OBSERVED FROM SHIP						
MATERIAL OBSERVATION SHEET						
DATE :				START TIME (UTC) :		
	TSG-ML	OSPAR	Quantify the objects according to the following size categories:	< 10 cm	10 cm <...< 50 cm	> 50 cm
Plastic	G2		Bags			
	G6	4	Bottles			
	G18	13	Crates and containers			
	G38		Cover / Packaging			
	G39		Gloves			
	G45		Mussels nets, Oyster nets			
	G48		Synthectic rope			

Fig. 2. Extract of The SeaCleaners observation sheet, material classification based on TSG-ML report

The nature of the waste contained in floating debris patch can be recorded via these surveys. Six major families of floating waste have been categorised by the TSG-ML (Technical Subgroup On Marine Litter under the Marine Strategy Framework Directive) and the OSPAR Commission [7]. The waste categories

---

[1] https://www.theseacleaners.org/en.

includes rubber, clothing/textiles, paper/cardboard, treated/worked wood, metals and plastic materials. Plastic wastes can also be classified in subcategories as follows (cf. Fig. 2), according to their nature plastic elements can not be treated in the same way.

3   Extraction Methodology and Data Management for Collecting Plastic Macro-waste at Sea

3.1   Challenges for Wastes Patches Tracking at Sea

Accordingly, several issues should be considered in relation with the respective data sources.

– *Data quality.* A first issue concerns data quality as incoming information could be incomplete or uncertain according to the data or observation source. For instance, for in situ information, the exact position of the waste could be imprecise but the observation can provide information concerning the waste patch composition. On contrary, data from satellites provides better information concerning the spatial extent of the waste patch but information for the composition could be incomplete. Those differences of quality necessitate fusion mechanisms to characterize a waste patch.
– *Data fusion.* A second issue concerns data comparison and fusion. Indeed, for a same macro-waste, different observations from different sources could have been done. However this implies to determine to which degree two given observations are related to the same waste patch by finding some similarities between observations that can be cross-related. Those similarities concern the spatial extent and the general composition of waste patch and to compare information in accordance with currents and meteorological conditions. The merging process depends on the quality of collected data (i.e., imagery process for remote sensor and observation condition for in situ observations). The registration of images acquired by different sensors is a challenging task, as it requires space and time registration and synchronisation. However many works have been developed for tracking object issues, methods used to tackle these problem are for instance the contour or any invariant feature detection or similarity measure relying on mutual information [21]. In the case of asynchronous time data collection, a prediction of movement is also needed.
– *Data and synopsis update.* This information fusion may be performed and extended using specific mechanisms to identify redundant or complementary information concerning one waste patch and update the database regularly. Some approaches so-called hybrid (i.e., merging results extracted from data stored in database with streaming data) could be useful [22], for continuously integrating new data in the database as well as updating synopses related to specific queries (i.e., density map of waste patch at sea).

3.2   A Data Model for Macro-waste Plastic Observation

As previously mentioned, waste patch geo-location data can be integrated by complementary protocols: from image data recorded by satellites, aircrafts and drones to in situ observations. As illustrated by Fig. 3, different complementary data sources are available for geo-locating macro-plastic areas (i.e., in situ visual observation, satellite or aerial images). Firstly, such data can be collected from the analysis of image data, using either satellite or aerial images from which, using analysis and filtering methods, the locations of plastic macro-waste can be extracted and approximated. As far as the location and associated semantic data are associated, the objective is to setup a visual and cartographic application that will provide basic elements (geo-location, surface of the macro-waste, approximate composition of marine litter) that can be even completed by some additional contextual data (e.g., photographs, comments). As waste patches are very likely to exhibit elements of different nature (e.g., clothes, paper, cardboard), documenting these compositions is also taken into account by the different composition tables identified by our model (i.e., other categories of marine litter exists, but we have only considered rubber and plastic elements for simplicity purposes). The following diagram describes the database model used for observing the distribution of marine litters and more specifically plastic macro-wastes at sea (Fig. 3).

For each observation, a polygon (or a point as a first consideration) that denotes a waste patch at a given observation time should be identified and recorded in the database. The extent and description observed for a given macro waste depends on the mean of observation. The model can associate two (or more) observations taken at different times when their their respective positions can be cross-related and if their characteristics provide sufficient similarities.

The data model for an assimilation of waste patch data has been defined and implemented within PostGIS. This favours further visualizations of waste patches according to their composition and spatial properties over time using appropriate queries. Those queries are more specifically density queries. For instance, a useful case is to have the location of all macro-wastes with a higher percentage of plastic and with a significant extent. For visualizations purposes, different layers can be considered according to the composition of the macro-wastes for instance.

3.3   Visualization of Geolocated Plastic Macro-wastes at Sea

The degree of plastic related to a marine litter, its drift/distance speed and weather conditions may influence the capability to recover an observed macro-waste. Indeed, marine litters with a strong plastic composition will be preferred because of the pyrolysis system on board the boat, marine litter moving away from the ship's range or present in unfavourable navigation areas will be neglected.

The implemented QGIS database supports queries and visualizations that can implicitly explain some emerging characteristics such as the nature of marine litters, their dimensions and drift according to currents.

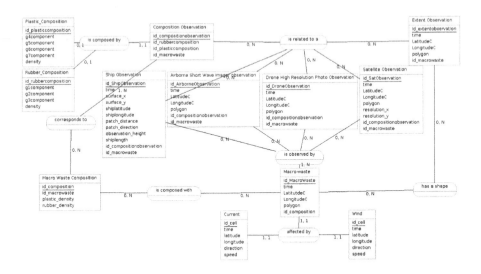

Fig. 3. Data model for macro-waste plastic observation at sea

Figure 4 illustrates the macro-waste present (in pink on the map) whose plastic concentration is relatively high. The white spaces represent macro-wastes in accordance with their spatial extent. The purple areas represent the areas for which the currents are low and for which the macro-waste present shouldn't move so much.

This map provides basic information on the distribution of these elements at sea, as well as their composition and surface area, this being useful for a first visualization of these wastes. However, the previous requests only supports visualization of the distribution of these macro-wastes without taking into account their drift, particularly in relation to meteorological and oceanographic elements. In the following part, a current drift estimation method is then proposed to generate a probability map at a time $t + \delta$.

4   Drift Prediction Model of Floating Debris Affected by Marine Currents

In this section two methods for evaluating marine currents are introduced. The first one is deterministic and the floating debris move along the current direction, while the second one discussed is stochastic to take into account the high variability of conditions at sea. The two methods are then applied and compared to determine the one which corresponds better to the movement of floating debris.

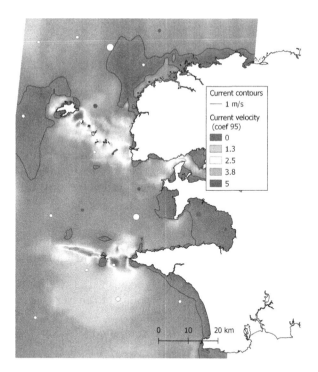

Fig. 4. Distribution of plastic macro-waste at sea near Brest

The objective is to propose a preliminary trajectory of floating waste patches in order to plan a navigation route to collect them. The main principles of the presented methods are to follow current direction. A similar method has been used in [20] where a simple estimation of particle displacement with time is derived from its last known position driven by the background ocean surface currents derived from the nearest data grid assuming negligible waves and wind-induced leeway effect. This method is itself similar to the conventional progressive vector diagram which has been used to estimate transport in the coastal ocean from velocity and time series [19]. These methods have been used in the context of search and rescue application, they provide a rapid response to the recovery of a missing person at sea. In this paper two methods to derive current from data with the same approach type are presented in the next subsections.

Definition 1: Floating waste patch. A patch A is defined as a polygon, its centroid is denoted C, localised by its latitude and longitude, LatC and LongC, and L and l the length and width of the waste patch.

As a first hypothesis, if the wasted patch is affected by marine currents, the drift is considered to be applied on its centroid.

Definition 2: Regular grid. A regular grid denoted G is used to define the region of interest with a size cell noted $Size_{cell}$. Each point of the grid, P_i, area the

location where the current are known for each time T_j taken at a regular time step, meaning $T_{j+1} - T_j$ is constant.

The $Size_{cell}$ value is chosen according to the phenomena to observe. In this case, the resolution of the grid is determined by current data provided by SHOM[2]. Data currents are denoted as $C(P_i, T_j)$, and they are composed by their velocities $V(P_i, T_j)$ and directions $A(P_i, T_j)$ at the points P_i and time T_j.

In the next section, $C(p, t)$ denotes the current which characteristics are needed. $P_{p/n}$ is denoted as the nth nearest points of p. In the case of a regular grid $n = 4$, actually the four points of the cell in which p is located. If the data is usually given according to a regular grid, the proposed method may be applied to different grid resolutions by modulating the number of nearest points. The distance between p to $P_{p/n}$ is denoted D_n.

4.1   Deterministic Process for Spatio-temporal Current Estimation

With this method the calculation of $C(p, t)$ begins with two spatial interpolations performed between the four current data points of the cell $P_{p/n=1..4}$ at the time T_j and T_{j+1} with $T_j < t < T + j$, Fig. 5.

The first spatial interpolation at T_j is computed as follows:

$$C(p, T_j) = \frac{1}{\sum_{i=1}^{n} 1/D_i} \cdot \sum_{i=1}^{n} \frac{1}{D_i} \cdot C(P_{p/i}, T_j) \text{ avec } i \in \{1, .., 4\}$$

A temporal interpolation is finally performed to compute $C(p, t)$ between the two spatial interpolations at $C(p, T_j)$ and $C(p, T_j + 1)$.

However, a linear interpolation may not describe the current flow correctly, particularly in shallow waters where the bathymetry for instance may largely affect the current characteristics from an observation data point to another. Instead, a stochastic method is proposed, where the floating waste is subject to be affected by a single current data point.

4.2   Stochastic Process for Spatio-Temporal Current Estimation

In this stochastic process, four temporal interpolations are made between T_j and T_{j+1} for the points $P_{p/n=1..4}$ in order to obtain the currents $C(P_{p/n=1..4}, t)$. Instead of performing a spatial interpolation between these four currents, the current values are affected accordingly to a single data current data point following a probability law:

$$Prob\{C(P_{p/k}, t\} = \frac{1}{\sum_{i=1}^{n} 1/D_i} \frac{1}{D_k} \text{ with } n = 4 \text{ and } k \in \{1, ., n\}$$

---

[2] https://www.shom.fr/.

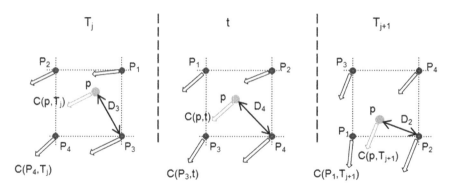

Fig. 5. Estimation of marine current at time t and location p

This method generates a corridor that denotes the possible presence of waste over a period time. The corridor is obtained by plotting a large number of possible waste trajectories, Fig. 6. This figure shows a set of probable trajectories for floating waste where trajectories have been derived by the method introduced above. This corridor shape of the trajectory can be explain by the location of the waste in open sea where the difference of current characteristic between two data point is very low (current vectors almost parallel). If the waste is located in coastal zone, in shallow water, near island where the current are affected by low bathymetry, trajectories be more diffused.

4.3   Intersection Estimation Between Ship and Floating Debris Trajectories

A first model of the ship behaviour and displacement is considered to model the collection of floating waste at sea. The boat direction to the waste is assumed as a straight line. This course is actually named ground track. The surface road is composed by the current drift and the ground track. The distance travelled by the ship, between its initial position and the waste, is calculated over a set of time intervals $\{T, T+1\}$, for each time interval the current velocity is estimated (see Sect. 4.2) to obtain the drift velocity due to the current. The real heading, the corrective course to steer sets by the navigator to compensate the drift and move towards the waste in a straight line, is also calculated, Fig. 7. Currents coming from the stern of the vessel will therefore tend to increase the navigation distance, while those coming from the bow will decrease ship velocity. A constant cruising speed is also assumed for the vessel velocity.

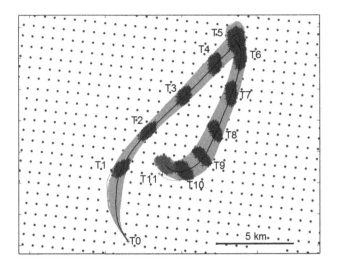

Fig. 6. Plot of 1000 trajectories with the 'stochastic' method in magenta, the green points represent the last position of waste after 12 h drift. The trajectory with the interpolation method is given in black. (Color figure online)

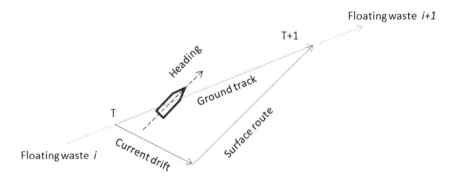

Fig. 7. Route followed by a ship subject to a marine current

The intersection calculation between the ship and the waste patches trajectories is performed according to the Algorithm 1. As a similar way to the displacement of the waste, with a time step calculation, the new position of the ship and the current characteristics affecting the vessel are estimated. As the vessel speed is significantly higher than the current velocity, the time step for calculating the vessel trajectory is lower than that the one used to calculate the floating debris.

Algorithm 1. Intersection computing

$W_i t$: Waste i position at t
Bt : Ship position at t
δt: time step
CT : last collect time
$t = 0$, $CT = 0$
while W_i are not collected **do**
    $t = t + \delta t$
    calculate $W_i t$
    for each W_i **do**
        calculate distance ship/waste: $D_{W_i t/Bt}$
    end for
    Select nearest waste: W_{near} with $D_{W_{near}/B} = min(D_{W_i t/Bt})$
    Calculate possible shipping distance towards W_{near} between t and CT: D_{nav}
    if $D_{W_{near}/B} = D_{nav}$ **then**
        Go collect W_{near}
        $Bt = W_{near} t$
        $CT = t$
    end if
end while

---

5 Application Case

To illustrate the behaviour of floating debris and ship trajectory models, as well as the recovery algorithm, a case study in the Iroise Sea, North West part of France, has been developed. Current data are provided by SHOM (Service Hydrographique et Océanographique de la Marine) around the Brest area in the Iroise Sea. The mesh size used is approximately 700 m in accordance with the file format provided by SHOM. One thousand waste drift simulations around Brest have been run using the stochastic method presented above. Depending on the distribution of the marine litter for each of the simulations, a density probability map was generated (see Fig. 8). Considering the number of times a macro-waste patch lies in a cell, a score is attributed to this cell. In the map, the color of the cells corresponds to the score of presence for a plastic macro-waste in order to visualize the probability of presence for plastic macro-waste.

The interest of such a map is multiple, it allows first of all to take into account the limit of our plastic drift model by integrating a random dimension. It then allows us to have an approximation of the distribution of waste, better than previous visualisations with QGIS which only allowed us to have a rendering of the marine litter at observation time. Finally, this map is the first essential element before setting up an algorithm to determine the best route for the boat to optimize the quantity of plastics collected.

In a second case, a ship navigating from the South must collect four waste patches of floating debris. This area is well known for its particular and strong tidal currents (particularly in the vicinity of the islands and islets).

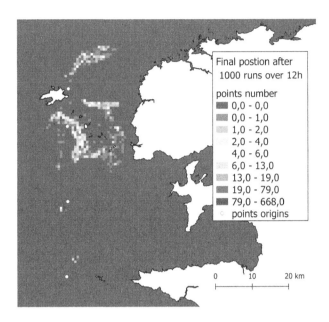

Fig. 8. Example of density map derived from four floating debris at sea (Color figure online)

Figure 9 shows the position of the waste at the initial time (red cross), their trajectory over a period of 12 h (black lines), the position of the vessel at initial time (green cross), its trajectory (green line) and respectively the collection points of waste (intersection of the trajectories of waste and the vessel) in green and magenta, for a vessel subject and not subject to currents. The tendency of the waste trajectory to turn back is due to the inversion of tidal currents every 6 h approximately in the region (semi-diurnal tide). By considering a navigation speed of 5 m/s, the waste patches are collected 2 h and 6 min, 4 h and 24 min, 7 h and 42 min and 9 h after the initial time.

The initial time is taken 6 h before high tide (beginning of the flow tide), the current is favourable to the ship, collecting the first two waste patches faster than a ship that would be not affected by current. About 6 h later, the current reverses (beginning of the ebb tide). The ship is mainly subject to counter flow, the waste collection times are delayed compared to a ship that would not be subject to current.

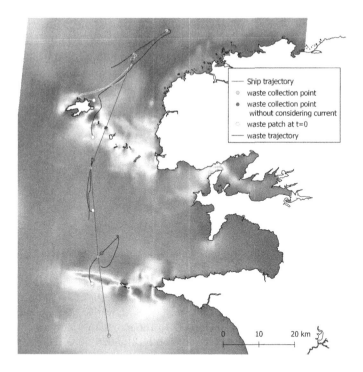

Ship trajectory
⊙ waste collection point
● waste collection point
   without considering current
○ waste patch at t=0
— waste trajectory

0   10   20 km

Fig. 9. Example of a ship trajectory for waste patch plastic collection (Color figure online)

6 Conclusion

An integrated approach for the observation and recovery of plastic waste in marine environments has been developed in this paper. The approach is based on an dual extraction mechanism and a database model that allows to monitor and visualize the movement of macro plastic at sea. Two methods of current estimation have then been proposed, one called deterministic, which evaluates the velocity and direction of the current by spatial and temporal interpolations, while another called stochastic is applied according to a probability law. Finally, a first model of the boat's behaviour for the collection of waste patch at sea has been proposed in order to build optimal routes taking into account the current and the reduced capacities of the boat used.

Future work may include additional data (e.g., winds, tides) to refine and improve the plastic garbage drift model both at the observation and recovery steps. This drift model take a specific place in the process of data merging that also need to be developed. Also the model of macro-waste trajectory following its centroïd for its displacement may find its limits if the macro-waste considered is too large. Another alternative could be to consider a macro-waste as a subdivision of macro-wastes when it is too large, so a macro-waste can be split into several parts or merged with others. Different weights according to the quality of the observation and mechanisms to determine the identification of the same patch for two different observations might be also applied. Finally, the characteristics of the vessel must be taken into account (as well as collaboration between different vessels), in particular to estimate the vessel's sailing time, as well as the current when the vessel moves in order to be able to route the vessel as efficiently as possible.

Acknowledgment. We would like to thanks Eric Le Plomb and Yannick Lerat of The SeaCleaners association for their discussions that greatly improved our understating of plastic debris issues in oceans.

References

1. UNEP: UNEP Year Book 2014: emerging issues in our global environment. United Nations Environment Programme (2014)
2. Eriksen, M., et al.: Plastic pollution in the world's oceans : more than 5 trillion plastic pieces weighing over 250,000 tons afloat at sea. Plos One **9**(12), 1–15 (2014)
3. Browne, M. A., Dissanayake, A., Galloway, T. S., Lowe, D. M., Thompson, R. C.: Ingested microscopic plastic translocates to the circulatory system of the mussel, Mytilus edulis (L.). Environ. Sci. Technol. **42**(13), 5026–5031 (2008)
4. Öztekin, A., Bat, L.: Microlitter pollution in sea water: a preliminary study from sinop sarikum coast of the southern black sea. Turk. J. Fish. Aquat. Sci. **17**(7), 1431–1440 (2017)
5. Rojas, J.: Plastic waste is exponentially filling our oceans, but where are the robots ? In: 2018 IEEE Region 10 Humanitarian Technology Conference (R10-HTC), pp. 1–6 (2018)
6. Cheshire, A., Adler, E., Barbière, J.: UNEP/IOC guidelines on survey and monitoring of marine litter. United Nations Environment, Programme/Intergovernmental Oceanographic Commission (2009)
7. Galgani, F., et al.: Guidance on monitoring of marine litter in European seas. In: European Commission (2013)
8. Lippiatt, S., Opfer, S., Arthur, C.: Marine debris monitoring and assessment: recommendations for monitoring debris trends in the marine environment. Marine Debris Monitoring and Assessment. NOAA Technical Memorandum NOS-OR&R-46 (2013)
9. Duis, K., Coors, A.: Microplastics in the aquatic and terrestrial environment: sources (with a specific focus on personal care products), fate and effects. Environ. Sci. Europe **28**(1), 1–25 (2016). https://doi.org/10.1186/s12302-015-0069-y
10. Sharma, S., Chatterjee, S.: Microplastic pollution, a threat to marine ecosystem and human health: a short review. Environ. Sci. Pollut. Res. **24**(27), 21530–21547 (2017). https://doi.org/10.1007/s11356-017-9910-8
11. Maximenko, N., Hafner, J., Niiler, P.: Pathways of marine debris derived from trajectories of Lagrangian drifters. Mar. Pollut. Bull. **65**, 51–62 (2012)
12. Law, K.L., et al.: Plastic accumulation in the North Atlantic subtropical gyre. Science **329**(5996), 1185–1188 (2010)
13. Eriksen, M., et al.: Plastic pollution in the South Pacific subtropical gyre. Mar. Pollut. Bull. **68**, 71–76 (2013)

14. Mace, T.H.: 2011: at-sea detection of marine debris: Overview of technologies, processes, issues, and options. Mar. Pollut. Bull. **65**, 23–27 (2012)
15. Jakovljević, G., Govedarica, M., Alvarez-Taboada F.: Remote sensing data in mapping plastics at surface water bodies. In: Conference: FIG Working Week 2019 - Geospatial Information for a Smarter Life and Environmental Resilience, Hanoi, Vietnam, 22–26 April 2019 (2019)
16. Goddijn-Murphy, L., Peters, S., Van Sebille, E., James, N.A., Gibb, S.: Concept for a hyperspectral remote sensing algorithm for floating marine macro plastics. Mar. Pollut. Bull. **126**, 255–262 (2018)
17. Kataoka, T., Murray, C.C., Isobe, A.: Quantification of marine macro-debris abundance around Vancouver Island, Canada, based on archived aerial photographs processed by projective transformation. Mar. Pollut. Bull. **132**, 44–51 (2018)
18. Moy, K., et al.: Mapping coastal marine debris using aerial imagery and spatial analysis. Mar. Pollut. Bull. **132**, 52–59 (2018)
19. Carlson, D.F., Muscarella, P.A., Gildor, H., Lipphardt Jr., B.L., Fredj, E.: How useful are progressive vector diagrams for studying coastal ocean transport ? Limnol. Oceanogr. Methods **8**(3), 98–106 (2010)
20. Shen, Y.-T., et al.: Applications of ocean currents data from high- frequency radars and current profilers to search and rescue missions around Taiwan. J. Oper. Oceanogr. **12**(2), 126–136 (2019)
21. Jinman, K., Gajera, K., Cohen, I., Medioni, G.: Detection and tracking of moving objects from overlapping EO and IR sensors. In: Conference on Computer Vision and Pattern Recognition Workshop (CVPRW), pp 123–123 (2004)
22. Salmon, L., Ray, C., Claramunt, C.: Continuous detection of black holes for moving objects at sea. In: 7th ACM SIGSPATIAL International Workshop on GeoStreaming, p. 2 (2016)

Research on Multi-scale Highway Risk Prediction and Assessment in China

Fengyun Mu$^{(\boxtimes)}$ (ID), Meng Yang (ID), and Mengmei Li (ID)

Chongqing Jiaotong University, Chongqing 400074, China
mfysd@cqjtu.edu.cn

Abstract. Highway flood risk assessment research is one of the current hotspots and difficulties in the field of natural disaster research, and it is also an important basis for disaster prevention and mitigation planning and management. The research analyzes the basic system of highway flood risk assessment from the aspects of basic concepts, influencing factors, and assessment contents of highway floods, and uses different methods to analyze the flood danger level from three dimensions: grid unit, small watershed unit, and town-level administrative unit. Combining qualitative analysis and quantitative assessment methods, the risk of highway floods at different spatial scales is evaluated, and the expected loss risk of highway floods in Banan District is estimated. The research shows that: (1) The grid can be used as a research unit to effectively explore the differences in the risk distribution within the study area due to its high data accuracy and complete sample information, it is suitable to use a neural network model for evaluation. (2) Based on the risk assessment at the small watershed scale, it is found that Banan District is mainly composed of low-risk and low-risk areas. The use of small watersheds as a research unit can reflect the spatial integrity of the impact factors of highway floods. The results It can provide decision support for river basin disaster management and prevention. (3) Based on the township-scale highway flood risk assessment, the township administrative unit is selected to analyze the flood danger levels of each town (street), which can meet the needs of management decision-making. Selecting different scales for risk assessment analysis can provide a scientific basis for the formulation, implementation, and allocation of disaster reduction resources for regional disaster management measures.

Keywords: Highway flood · Harzard · Multiscale · Risk assessment · Banan district

1 Introduction

With the intensification of global climate change, extreme weather around the world is frequent, and the number of natural disasters caused by extreme weather has increased. Highway floods are a series of natural disasters caused by extreme rainfall and heavy rain. Highway flood research can help China's socialist construction, ensure the implementation of the "Belt and Road" and "Yangtze River Economic Belt" strategies, and

© Springer Nature Switzerland AG 2020
S. Di Martino et al. (Eds.): W2GIS 2020, LNCS 12473, pp. 230–240, 2020.
https://doi.org/10.1007/978-3-030-60952-8_22

provide theoretical support for highway flood research. In recent years, due to the combined effects of extreme catastrophic weather and economic and social activities, natural disasters such as floods, mudslides, and landslides have occurred frequently, causing casualties, property damage, infrastructure damage, and environmental damage [1–4]. According to statistics from the International Disaster Database EM-DAJ, from 2000 to 2017, a total of 235 large-scale floods occurred in China. The research on highway floods aims to clarify the formation mechanism of highway floods, provide scientific basis for the early warning, management, risk assessment and disaster assessment of highway floods, and also have important reference value for the planning and development of regional planning, traffic planning and urban planning.

However, How to select practical, pertinent and representative factors and scientific evaluation methods to evaluate highway floods is also a crucial issue. Hanse (1984) et al. [5] combined field surveys and remote sensing image interpretation methods, and used gradient factors and mutation factors as the influencing factors to construct disaster data. Based on GIS and flood disaster site cataloging, the flood disaster risk and activity level were evaluated in mountainous areas. And constantly analyze the cause of the error, improve the data processing method, and finally realize the assessment of mountainous flood danger areas. Qi (2014) et al. [6], Lin (2015) et al. [7] constructed a county highway flood risk assessment index system from the aspects of topography, rainfall, lithology, river network and vegetation, etc., and used the comprehensive index method to establish a comprehensive index model for highway flood risk assessment. Although various scholars have begun to conduct pilot research on flood prevention and control planning in some typical cities and towns, they have not paid enough attention to floods on mountain highways as key lifeline projects, and they only involve the highway water damage mechanism and water damage prevention and zoning.

In view of the lack of in-depth study on risk assessment of meso scale and above, carrying out the study on road flood risk assessment and its spatial distribution pattern is not only the urgent requirement to change the concept of disaster prevention and mitigation in China, effectively reduce the loss of road flood, but also can fill the shortage of current road flood risk assessment study.

2 Research Train of Thought

Flood disaster is a complex disaster system which is affected by the interaction of human and nature [8–14]. It has natural and social attributes. Its occurrence and risk are closely related to the local natural environment and human factors. There are four conditions for the formation of highway flood disaster: ① disaster causing factors, i.e. triggering factors, which generally include rainstorm, dam break, reservoir flood, etc. ② disaster pregnant environment, i.e. the environment of highway flood disaster, generally including underlying surface environment and meteorological climate, etc. ③ disaster bearing body, i.e. human society directly or indirectly affected and damaged by highway flood disaster The main body of the meeting generally includes highway itself, highway auxiliary facilities, transportation industry and human beings, etc. ④ disaster situation, in the environment of highway flood disaster, the direct or indirect casualties, economic losses and ecological environment damage caused by the disaster causing factors acting on the disaster bearing body (Fig. 1).

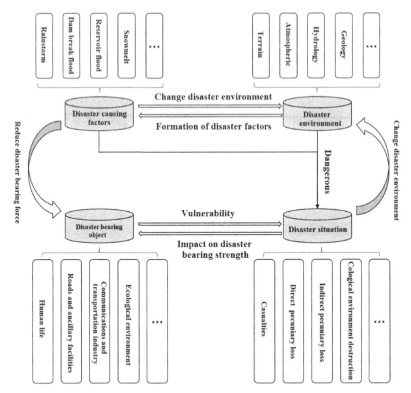

Fig. 1. Schematic diagram of highway flood disaster system

3   Research Methods and Basic Process

In the flood risk assessment, the location and influence range of the flood disaster are closely related to the geographical and spatial characteristics, and less related to the artificial administrative division [14–17]. Based on the grid scale, the highway flood risk assessment can quantify a variety of data spaces to a unified and easy to analyze scale, and realize the spatial refinement of data attributes, which provides a new way for the expression of traditional data, breaks through the constraints of administrative boundaries, and effectively explores the differences in risk distribution within administrative divisions. In this paper, the basic evaluation unit of highway flood risk assessment is determined by grid division of the study area, and the data is quantized to each evaluation unit by using different spatial data fine quantization methods, and the grid risk assessment model is constructed by using neural network method to realize the grid based highway flood risk assessment.

Accurate and effective construction of assessment methods should be targeted, which has a very important impact on the results of disaster risk assessment This study will discuss the risk assessment methods and related models of highway flood disaster from the meaning of risk assessment [17–19]. Risk assessment is based on risk assessment and vulnerability assessment. These methods and models continue to improve the level of disaster risk analysis. These models have their own advantages and disadvantages. In practical application, they are selected and applied according to the characteristics and suitability of the specific analysis model (Fig. 2).

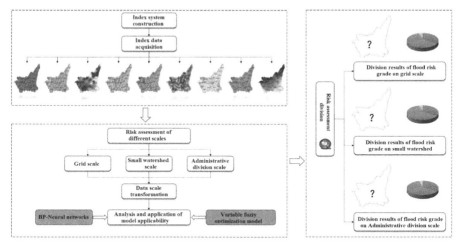

Fig. 2. Flow chart of multi-scale highway flood risk assessment method

4   Analysis of Research Results

4.1   Risk Assessment Based on Grid Scale

Based on the grid scale, the highway flood risk assessment can quantify a variety of data spaces to a unified and easy to analyze scale, and realize the spatial refinement of data attributes, which provides a new way for the expression of traditional data, breaks through the constraints of administrative boundaries, and effectively explores the differences in risk distribution within administrative divisions (Fig. 3).

According to the results of flood hazard zoning in Banan District Based on grid scale, the flood hazard level in Banan District is mainly micro hazard and low hazard. From the perspective of spatial distribution, the micro and low risk areas are mainly concentrated in the south of Banan District. The rainfall intensity in the south of Banan District is significantly lower than that in the West and North near the Yangtze River, and the geomorphic type is mostly the ridge and the upper part of the hillside, with little rain erosion (Table 1).

4.2   Risk Assessment Based on Small Watershed Scale

Risk is the natural attribute of disaster. It is generally used to describe the distribution probability of natural environment such as regional disaster pregnant environment and disaster causing factors. From the definition of risk, it can be seen that risk assessment

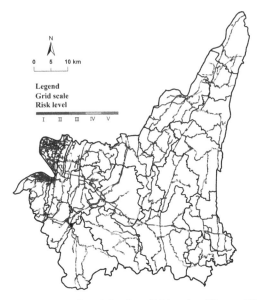

Fig. 3. Risk assessment of road flood on Grid scale of Banan District Based

Table 1. Area proportion of each danger level area in Banan District

Proportion of hazard	Class	Area (km^2) (%)	
Microhazard I	564.01	30.75	
Low risk II	825.44	45.00	
Moderate risk III	337.82	18.42	
High risk IV	106.96	5.83	

is a kind of assessment on the natural environment of disaster pregnant environment, while for the terrain characteristics of southwest mountain and hills, small watershed is a natural unit with complete natural ecological process, and is the basis and key of regional disaster comprehensive management (Table 2 and Fig. 4).

Table 2. Area proportion of small watershed risk grade area in Banan District

Proportion of hazard	Class	Area (km^2) (%)	
Microhazard I	637.92	34.78	
Low risk II	622.86	33.96	
Moderate risk III	496.13	27.05	
High risk IV	77.31	4.21	

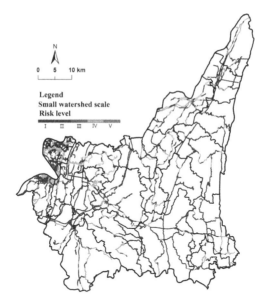

Fig. 4. Risk assessment of road flood on small watershed scale of Banan District Based

4.3 Risk Assessment Based on Town Scale

The evaluation process based on grid scale fully considers the non-uniformity of regional natural conditions distribution, and its evaluation results are more accurate. The evaluation based on small watershed scale is based on the analysis and evaluation of a complete natural ecological process on the natural unit, which retains the integrity of the disaster environment and data consistency, and more complex definition of risk evaluation. From the perspective of man-made disaster management, the evaluation based on administrative unit is more effective and more practical than that based on natural unit. Based on the town scale risk assessment, it is convenient for decision makers and managers to carry out road flood prevention, management and management, so that the corresponding disaster prevention and management measures are more targeted, so as to ensure the reasonable distribution of limited resources (Table 3 and Fig. 5).

4.4 Multi Scale Risk Assessment of Highway Flood

According to the survey width of 2 km on both sides of the road, 217 historical flood disaster data are extracted as observation points, and the disaster risk prevention and control level set in the disaster data information table is taken as the observation value. The evaluation accuracy is verified by comparing the error between the evaluation value and the observation value. It is verified that the overall accuracy of the superposition operation based on grid, small watershed and town scale is 0.736, 0.675 and 0.608, respectively. In terms of the accuracy of the simulation results, all of them have a certain degree of credibility. The overall high accuracy is the calculation results based on the grid scale, followed by the calculation results based on the small watershed scale. These

Table 3. Flood hazard level of towns (Streets) in Banan District

Town (Street)	Proportion of hazard	Town (Street)	Proportion of hazard
Lijiatuo street	high risk IV	Ersheng town	low risk II
Huaxi street	high risk IV	tianxingsi town	low risk II
Yudong Street	medium risk III	Jielong town	micro risk I
Nanquan Street	low risk II	Mudong town	medium risk III
Longzhouwan street	high risk IV	Jiangjia town	low risk II
Yipin Street	micro riskI	Maliuzui town	moderate risk III
Jieshi town	low risk II	Shuanghekou town	low risk II
Anlan town	low risk II	Fengsheng town	moderate risk III
Huimin street	iModerate risk III	Dongwenquan town	low risk II
Nanpeng Street	medium risk III	Shilong Town	micro risk I
Shengdengshan town	micro risk I	Shitan town	micro risk I

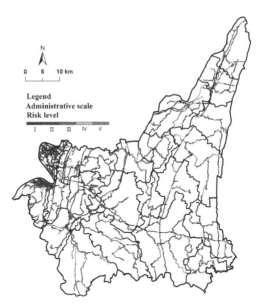

Fig. 5. Risk assessment of road flood on administrative scale of Banan District Based

two scales are suitable for the research unit of risk assessment and can get more accurate results. However, in practical application, the superposition results based on town scale contain the attribute information of town administrative divisions, which is convenient for managers to query the responsibility information of the highway. Therefore, according to the different purposes of the assessment, the suitability of different scales should be analyzed, and the best scale should be selected as the assessment unit of flood risk assessment.

The evaluation results with the highest simulation accuracy are selected for discussion. Among the superposition results based on grid scale, the most low-risk highways are 620.8 km long, accounting for 39.8% of the total length of the study highway; the medium-risk highways are 505.3 km long, accounting for 32.4% of the total length of the study highway, of which 11.9% are class IIB, that is to say, the highways with high risk become class II Risk highways, and 11.5% are class IIC, that is, the highways with high vulnerability are formed It is a class II Risk highway; the length of the high-risk highway is 345.7 km, accounting for 22.2% of the total length of the research highway, of which 11.2% are class III B. these highways are of high risk because of their high probability of occurrence of flood, but their vulnerability level is not high. 8.7% of the highways are of high vulnerability, but the probability of occurrence of flood is not high, and they are also of high risk; the length of the high-risk highway is 88 km, accounting for 5.6% of the total length of the study road, mainly distributed in lijiatuo street, North Yudong street, South NANPENG street, middle Jieshi Town, North Mudong Town, West Shuanghekou town and West maliuziao town.

The high-risk highway of middle highway flood in Banan District is about 29.16 km long, accounting for 19.4% of the expressway, including the east section and Jieshi section of G65 Baotou Maoming expressway, and the boundary stone section of G75 Lanhai expressway, etc.; there are many high-risk sections of Shanghai Chongqing south highway, such as guojiaping Village section of Mudong Town, huangshakan village to badouchong Village section of maliuzui Town, etc. In the national highway, there are basically medium-risk highway and high-risk highway, and the high-risk highway accounts for less 13.6% of the total length of the national highway, including Jianhe road in Yudong street, yipinzheng street in Yipin town and Gaoping Village section of G210, which are 3.76 km high-risk highway in total. The medium and high-risk highway of the provincial highway is about 35.26 km long, accounting for 15.79% of the total length of the provincial highway, among which there are many high-risk highways in S106 of Yudong street; the south section of Yunan Avenue is basically high-risk highway; s415 is high-risk highway in Guihuaping village of Jiangjia Town, and about 1.9 km long high-risk highway continuously appears between dengjiawan village and longdongbang village; 3.4K long highway in mengziwan village of S103 of Mudong town There are about 1.2 km high-risk roads at the junction of Shuanghekou town and maliuzui town of S103. The county roads in Banan District are mainly medium-risk roads, accounting for 60.4%, followed by high-risk roads, accounting for about 6.85%. The high-risk roads account for a small proportion, about 1.1 km, only 0.9% of the total length of the county road. They are basically distributed in Huaxi, Yudong and Mudong towns. There are many high-risk roads in X235, X238, x760, x761, x765 and x769 county roads. The flood risk level of township and village roads and other low-grade roads is relatively good, and the roads with higher risk and above account for about 4.5% of the total length of such roads, among which, there are more continuous and high-risk roads in Banan Binjiang Road, Yudong Jiangbin Road, Anlan town Y117 and Mudong town c535. For the above-mentioned roads, the regional disaster prevention and control management department shall formulate appropriate comprehensive prevention and control measures to reduce the road flood risk according to the flood risk and the road vulnerability.

5   Conclusion and Discussion

Risk assessment and zoning research can not only provide decision support for disaster prevention and mitigation, but also provide important reference for urban construction and rational use of environmental resources. Based on the definition of flood hazard and considering the factors affecting the risk of highway flood, this study selects 9 indexes, i.e. terrain slope type, geological lithology, average precipitation from April to October, historical development of geological hazard, vegetation coverage, river network density, soil runoff curve number, human activity intensity index and rainstorm intensity, from the two aspects of disaster pregnant environment and disaster causing factors, and adopts different models Methods: the risk level of flood disaster in Banan District was evaluated with a grid of 3030 m, a small watershed and a town as the smallest research units.

(1) Based on the grid scale of highway flood risk assessment, through the precise quantification of data, the relatively complete assessment information is retained, which makes the assessment results more accurate and reliable, and can effectively explore the differences in risk distribution within the administrative divisions. By using BP neural network model, the distribution of flood risk grade in Banan District is obtained. The results show that the main areas in Banan District are micro risk area and low risk area, accounting for 30.75% and 45% of the total area respectively, and the medium risk area and high risk area accounting for 18.42% and 5.83% of the total area respectively. The accuracy of the model is 0.805 by k-fold method.

(2) Based on the small watershed scale of highway flood risk assessment, taking natural unit as the assessment unit can better reflect the integrity of highway flood impact factors, and provide decision support for the basin disaster management and prevention work. Using the variable fuzzy optimization model to calculate the flood risk level of each small watershed in Banan District, the results show that the small watershed in Banan District is mainly micro risk and low risk, accounting for 34.78% and 33.96% of the total area of the county, respectively, and the medium risk and high risk areas accounting for 27.05% and 4.21% of the total area. The accuracy of the model is 0.763 verified by the full number test method.

(3) The risk assessment of highway flood based on town scale can meet the needs of management decision-making. The comprehensive risk degree of administrative units at all levels is analyzed from a macro perspective, which provides a scientific basis for the formulation and implementation of disaster management measures and the allocation of disaster reduction resources in Banan District. Using the variable fuzzy optimization model to calculate the flood risk level of towns (Sub district office) in Banan District, the results show that there are three high-risk towns (Sub district office) in Banan District, the majority of which are low-risk and medium-risk towns (Sub district office). The relevant departments should strengthen the disaster prevention efforts of medium and high-risk towns, and check the potential disaster points, so as to reduce the losses caused by flood.

Highway flood risk assessment is a complex system engineering, which involves the knowledge of meteorology, geology, hydrology and other disciplines. At present, a lot of research and application achievements have been made in the research of flood risk in China, but there are still many problems to be further discussed in the vulnerability and risk assessment of highway flood. Combining the actual situation of road flood in Banan District, this paper explores the method and technology of road flood risk assessment, and obtains some research results, but there are still some deficiencies to be further improved and explored in the follow-up study.

References

1. Cai, H., Rasdorf, W., Tilley, C.: Approach to determine extent and depth of highway flooding. J. Infrastructure Syst. **13**(2), 157–162 (2007)
2. Seejata, K., Yodying, A., Wongthadam, T., et al.: Assessment of flood hazard areas using analytical hierarchy process over the lower yom basin, sukhothai province. Procedia Eng. **212**, 340–347 (2018)
3. Arabameri, A., Rezaei, K., CerdàA, A., et al.: A comparison of statistical methods and multi-criteria decision making to map flood hazard susceptibility in Northern Iran. Sci. Total Environ. **2019**(660), 443–458 (2019)
4. Qin, Q., Lin, X., Tang, H., et al.: Environmental zoning of road flood in Wanzhou District, Chongqing. Chongqing Jiaotong Univ. Nat. Sci. Edition **30**(1), 89–94 (2011)
5. Hanse, A.: Land slide hazard analysis. In: Brusden, D., Prior, D.B.: Slope Instability, pp. 523–543 (1984)
6. Qi, H.L., Tian, W.P., Wang, D., et al.: Risk assessment index of highway flood disaster. Disaster Sci. **29**(3), 44–47 (2014)
7. Lin, X.S., Chen, H.K., Wang, X.J., et al.: Environmental zoning of highway flooding in Southwest China. Resources and Environ. Yangtze River Basin **21**(2), 251–256 (2012)
8. Cutter, S.L., Scott, M.M.S.: Revealing the vulnerability of people and places: a case study of Georgetown County, South Carolina. Ann. Assoc. Am. Geographers **90**(4), 713–737 (2000)
9. Rufat, S., Tate, E., Burton, C.G., et al.: Social vulnerability to floods: review of case studies and implications for measurement. Int. J. Disaster Risk Reduct. **14**, 470–486 (2015)
10. Wang, Z., Lai, C., Chen, X., et al.: Flood hazard risk assessment model based on random forest. J. Hydrol. **527**, 1130–1141 (2015)
11. Elkhrachy, Ismail: Flash flood hazard mapping using satellite images and gis tools: a case study of Najran City, Kingdom of Saudi Arabia (KSA). Egyptian J. Remote Sens. Space Sci. **18**(2), S1110982315000307 (2015)
12. Teng, J., Jakeman, A.J., Vaze, J., et al.: Flood inundation modelling: a review of methods, recent advances and uncertainty analysis. Environ. Modell. Software **90**, 201–216 (2017)
13. Nobre, A.D., Cuartas, L.A., Momo, M.R., et al.: HAND contour: a new proxy predictor of inundation extent. Hydrol. Processes **30**(2), 320–333 (2016)
14. Ghimire, B., Chen, A.S., Guidolin, M., et al.: Formulation of a fast 2D urban pluvial flood model using a cellular automata approach. J. Hydroinform. **15**, 676 (2013)
15. Diermanse, F.L.M., Geerse, C.P.M.: Correlation models in flood risk analysis. Reliability Eng. Syst. Safety **105**, 64–72 (2012)
16. Ashraf Mohamed Elmoustafa: Weighted normalized risk factor for floods risk assessment. Ain Shams Eng. J. **3**(4), 327–332 (2012)
17. Young-Oh, K., Seung Beom, S., Ock-Jae, J.: Flood risk assessment using regional regression analysis. Nat. Hazards, **63**(2), 1203–1217 (2012)

18. Alfieri, L., Feyen, L., Dottori, F.: Ensemble flood risk assessment in europe under high end climate scenarios. Global Environ. Change **2015**(35), 199–212 (2015)
19. Abuzied, S., Yuan, M., Ibrahim, S.: Geospatial risk assessment of flash floods in nuweiba area, Egypt. J. Arid Environ. **133**, 54–72 (2016)

Towards Faster Space-Efficient Shortest Path Queries (Work-in-Progress)

Stefan Funke[(✉)]

University of Stuttgart, Stuttgart, Germany
funke@fmi.uni-stuttgart.de

Abstract. We report on the current development of a shortest path computation technique in road networks. While not as fast as the currently fastest schemes, our new approach is very simple and exhibits competitive times with only moderate space overhead.

Keywords: Route planning · Speed-up techniques

1 Introduction

The last 10–15 years have seen tremendous progress when it comes to the problem of efficiently computing shortest paths in *real-world road networks*. Here the main idea is to spend some time in a preprocessing step where auxiliary information about the network is gathered, such that subsequent queries can be answered much faster than via standard Dijkstra's algorithm. One might classify most of the employed techniques into two classes: ones based on *pruned graph search* and ones based on *distance lookups*. Most approaches fall into the former class, e.g., reach [6], arc-flags [4], or contraction hierarchies (CH) [5]. Here, basically Dijkstra's algorithm is given a hand to ignore some vertices or edges during the search. The achievable speed-up compared to plain Dijkstra's algorithm ranges from one magnitude ([6]) up to three orders of magnitude ([5]). In practice, this means that a query on a country-sized network can be answered in less than a *millisecond* compared to few seconds of Dijkstra's algorithm. While these methods directly yield the actual shortest path, the latter class is primarily concerned with the computation of the (exact) distance only – recovering the actual path often requires some additional effort. Examples for such distance-lookup-based methods are transit nodes [3] and hub labels [1]. They allow for the answering of distance queries another one or two orders of magnitude faster. Yet, methods based on pruned graph search are more popular in practice, due to their low space overhead. For example, for a network of around 20 million nodes, the scheme in [1] requires to store for each node in the order of hundreds distance labels. See [2] for a comprehensive survey on the topic.

1.1 Contribution and Outline

In this work-in-progress-report we propose a hybrid acceleration technique that combines contraction hierarchies and hub labels. With appropriate choices of

S. Di Martino et al. (Eds.): W2GIS 2020, LNCS 12473, pp. 241–244, 2020.
https://doi.org/10.1007/978-3-030-60952-8_23

the parameters, our hybrid approach only requires additional space comparable to the size of the road network itself, yet improves query times by one order of magnitude. We first recapitulate basics of the contraction hierarchy (CH) scheme as well as the CH-based hierarchical hub labelling scheme, and then show how to combine HL and CH to obtain a fast, yet space efficient query scheme. We conclude with a brief experimental evaluation on real world data.

2   Preliminaries: Contraction Hierarchies and Hub Labels

The *contraction hierarchies* approach [5] computes an overlay graph in which shortcut edges span shortest paths. This reduces the hop length of optimal paths and therefore allows a variant of Dijkstra's algorithm to answer queries more efficiently. The preprocessing is based on a *node contraction* operation. Here, a node v and its incident edges are removed from the graph. In order not to affect shortest path distances between the remaining nodes, shortcut edges are inserted between all neighbors u, w of v, if and only if uvw was a shortest path. The cost of the new shortcut edge (u, w) is set to the summed costs of (u, v) and (v, w). In the preprocessing phase all nodes are contracted one-by-one in some order. The rank of the node in this contraction order is also called the *level* of the node. Note that we can also allow the simultaneous contraction of non-adjacent nodes, hence many nodes might have the same level.

Having contracted all nodes, a new graph $G^+(V, E^+)$ is constructed, containing all original edges of G as well as all shortcuts that were inserted in the contraction process. An edge $e = (v, w)$ – original or shortcut – is called upwards, if the level of v is smaller than the level of w, and downwards otherwise. By construction, the following property holds: For every pair of nodes $s, t \in V$, there exists a shortest path in G^+, which first only consists of upwards edges, and then exclusively of downwards edges. This allows employ a bidirectional Dijkstra only considering upwards edges in the search starting at s, and only downwards edges in the reverse search starting in t. This reduces the search space significantly and allows for answering of shortest path queries within the *milliseconds* range compared to *seconds* on a country-sized road network.

Hub Labelling is a scheme to answer shortest path distance queries which differs fundamentally from graph search based methods. Here the idea is to compute for every $v \in V$ a *label* $L(v)$ such that for given $s, t \in V$ the distance between s and t can be determined by just inspecting the labels $L(s)$ and $L(t)$. All the labels are computed in a preprocessing step (based on the graph G), later on, the graph G can even be thrown away. There have been different approaches to compute such labels (even in theory); we will be concerned with labels that work well for road networks and are based on CH again, following the ideas in [1]. To be more concrete, the labels we are constructing have the following form:

$$L(v) = \{(w, d(v, w)) : w \in H(v)\}$$

We call $H(v)$ a set of *hubs* – important nodes – for v. The hubs should be chosen such that for any s and t, the shortest path from s to t intersects $L(s) \cap L(t)$.

If such label sets could be computed, the computation of the shortest path distance between s and t boils down to determining the node $w \in L(s) \cap L(t)$ minimizing the summed distance. If the labels $L(.)$ are stored lexicographically sorted, this can be done in a very cache-efficient manner in time $O(|L(s)| + |L(t)|)$.

Given a CH, there is an easy way of computing such labels: run an upward Dijkstra from each node v and let the label $L(v)$ be the settled nodes with their respective distances. Clearly, this yields valid labels since CH answers queries exactly. The drawback is that the space requirement is quite large; depending on the metric and the CH construction, one can expect labels consisting of several hundreds to thousands node-distance pairs. It turns out, though, that many of the labels created in such a manner are useless as they do not represent shortest-path distance (as we restricted ourselves to a search in the upgraph only); pruning out those reduces the number of labels by a factor of 4. A source target distance query can then be answered in the *microseconds* range.

3   Combining HL and CH

The idea of our hybrid approach is very simple and motivated by the naive way of constructing hub labels based on an already precomputed contraction hierarchy; in principle we could simply start from every node v an upwards Dijkstra and store the visited nodes with their respective distances. Nobody constructs hub labels this way (see [1] how it is done in practice), yet this intuition is very helpful for our hybrid approach. The key observation is the following: if we have nodes u and u', and u' is in the upgraph of u at distance $d(u, u')$ (that is, reachable via an upward Dijkstra from u), then the upgraph of u' is a subgraph of the upgraph of u. This means in particular, having a label for node u' we could derive (part of) the label for u by adding $d(u, u')$ to all distances in the label for u'.

Now assume we have constructed and stored hub labels for all nodes of level at least, let's say $L_{hl} = 4$ and we are to answer a query from node u to v with the levels of source and target being smaller than L_{hl}. The idea is to use the precomputed hub labels above L_{hl} to quickly construct the actual hub labels for u and v and then answer the query as usual. This can be done as follows: we start an upgraph Dijkstra (as in the regular CH query) from u but ignoring outgoing edges of nodes of level at least L_{hl}. If L_{hl} is rather small (e.g., less than 16), this search will die out very quickly. Yet it provides us with distances to all nodes above level L_{hl} that are reached first from u in the upgraph Dijkstra. For all these nodes we collect the precomputed labels and add the respective distances (as computed during the upgraph Dijkstra) to their distance labels. The union of all these (hub, distance) pairs is the label for u. We proceed accordingly for v and use the two constructed labels to answer the query. If the highest-level node on the shortest path from u to v has level below L_{hl} we determine the distance between u and v already after the two upgraph searches.

4   Implementation and Experimental Evaluation

Our hybrid scheme was implemented in C++ and executed on a single core of an AMD Ryzen 1800X running Ubuntu 18.04 for the road network of the federal state of Baden-Württemberg with 3.97 million nodes and around 8 million edges from the OpenStreetMap project. With added CH shortcuts we have a graph with around 14.56 million edges consuming around 700 MB of space. CH as well as CH-based hub labels were computed using the standard approaches in [1,5]. Levels start at 0, the maximum level is 165. The important figures for our scheme are space requirement and query time. Setting the minimum level L_{hl} determines how to trade between contraction hierarchy and hub labels. For $L_{hl} = 0$, we get pure hub labels, for $L_{hl} = \infty$, we obtain the original contraction hierarchy scheme. In Table 1 we have characterized the trade-off between query time and space consumption for the data set of Baden-Württemberg.

Table 1. Trade-off between average query time and space consumption.

L_{hl}	0	1	2	4	8	16	32
Space (HL down to L_{hl})	6,959 MB	3,802 MB	2138 MB	849 MB	261 MB	50 MB	6 MB
Query time Dijkstra (in μs)	859,942						
Query time CH (in μs)	171						
Query time HL (in μs)	1.35						
Query time hybrid (in μs)	5.04	7.51	8.96	11.06	15.34	31.96	96.95
# labelled nodes collected	2	3	5	7	12	29	85

If we set $L_{hl} = 4$, i.e., storing hub labels only for nodes of level at least 4, we incure a space overhead of only 849MB. Yet, we achieve average query times almost factor 80 better than plain CH. Query times are about a factor of 8 worse than pure hub labels, but the latter require 7GB storage.

References

1. Abraham, I., Delling, D., Goldberg, A.V., Werneck, R.F.: Hierarchical hub labelings for shortest paths. In: Epstein, L., Ferragina, P. (eds.) ESA 2012. LNCS, vol. 7501, pp. 24–35. Springer, Heidelberg (2012). https://doi.org/10.1007/978-3-642-33090-2_4
2. Bast, H., et al.: Route planning in transportation networks. CoRR, abs/1504.05140 (2015)
3. Bast, H., Funke, S., Sanders, P., Schultes, D.: Fast routing in road networks with transit nodes. Science **316**(5824), 566–566 (2007)
4. Bauer, R., Delling, D.: SHARC: fast and robust unidirectional routing. In: ALENEX, pp. 13–26. SIAM (2008)
5. Geisberger, R., Sanders, P., Schultes, D., Vetter, C.: Exact routing in large road networks using contraction hierarchies. Transp. Sci. **46**(3), 388–404 (2012)
6. Gutman, R.J.: Reach-based routing: a new approach to shortest path algorithms optimized for road networks. In: ALENEX/ANALCO, pp. 100–111. SIAM (2004)

Author Index

Printed in the United States
By Bookmasters